Beihang Series in Space Technology Applications
北京航空航天大学"空间技术应用"系列丛书

Guidance Principle of Missiles
导弹制导原理

Jiang Jiahe
江加和

北京航空航天大学出版社

Abstract

The intent of this book is to present guidance and control principle of tactical missiles. It includes basic concepts of guided missile, fundamental concepts of vehicle dynamics, dynamical equations and kinematic equations of vehicle, longitudinal state equation and transfer functions, lateral state equation and transfer functions, fundamental principle of missile guidance and control system, guidance laws, autopilot design, command guidance systems, homing guidance systems, and guidance and control system hardware-in-the-loop simulation.

This book is suitable for international postgraduate and advanced undergraduates majoring in navigation, guidance and control, and also suitable for engineering and technical personnel engaged in the design and development of guided missiles.

图书在版编目(CIP)数据

导弹制导原理 = Guidance Principle of Missiles：英文 / 江加和编著. -- 北京：北京航空航天大学出版社，2012.6
ISBN 978-7-5124-0752-7

Ⅰ.①导… Ⅱ.①江… Ⅲ.①导弹制导－英文 Ⅳ.①TJ765.3

中国版本图书馆 CIP 数据核字(2012)第 045313 号

版权所有，侵权必究。

Guidance Principle of Missiles
导弹制导原理
Jiang Jiahe
江加和
责任编辑　宋淑娟

*

北京航空航天大学出版社出版发行

北京市海淀区学院路 37 号(邮编 100191)　http://www.buaapress.com.cn
发行部电话：(010)82317024　传真：(010)82328026
读者信箱：bhpress@263.net　邮购电话：(010)82316936
北京时代华都印刷有限公司印装　各地书店经销

*

开本：787×960　1/16　印张：20.25　字数：454 千字
2012 年 6 月第 1 版　2012 年 6 月第 1 次印刷　印数：1 500 册
ISBN 978-7-5124-0752-7　定价：69.00 元

若本书有倒页、脱页、缺页等印装质量问题，请与本社发行部联系调换。联系电话：(010)82317024

Preface

The intent of this book is to present guidance and control principle of tactical missiles. The book describes onboard control and guidance devices and off-board guidance devices, and moreover, discusses the analysis and design about the system. This book consists of eleven chapters. Chapter 1 introduces basic concepts of guided missile. Chapter 2 presents the fundamental concepts of vehicle dynamics. The forces and moments acting on the vehicle are also presented. Chapter 3 discusses vehicle dynamical equations and kinematic equations. Based on small-disturbance theory, Chapter 4 discusses vehicle longitudinal state equations and transfer functions. Chapter 5 deals with lateral state equations and transfer functions. Chapter 6 presents fundamental principle of missile guidance and control system. Chapter 7 focuses on guidance laws. Autopilot design is discussed in Chapter 8. Chapter 9 is devoted to command guidance systems. Chapter 10 focuses on homing guidance system. Chapter 11 summarizes guidance and control system hardware-in-the-loop simulation.

To help readers understand the concepts presented in the text, a number of worked-out examples are given throughout the book. MATLAB source code listings about these examples are also presented in the text. The examples demonstrate ideas in analysis, computer-aided design, simulation, and numerical algorithms. The examples, which are easily reproducible, offer reference for students to study or for engineers to perform engineering design. This book differs from similar books on the subject in that it presents a detailed account of vehicle aerodynamics, vehicle mathematical models, flight control, and guidance devices (mainly including radar guidance devices and infrared devices). The book maintains certain essential contents about guidance theory. Further more, it discusses guidance techniques from engineering application view. Guidance devices and guidance methods directly or indirectly come from actual missile types. Hardware-in-the-loop simulation is a vital phase in development process for guided missiles. This technique is described in the book as well.

This book is intended as a textbook for the guidance principle course for international postgraduates, and will be of benefit to advanced undergraduates majoring in navigation, guidance and control as well. The book offers very useful help to engineers engaged in the design and development of guided missiles.

I would like to take the opportunity to thank many people who contributed in some way to the contents and publication of this book. First of all, I would like to express my appreciation to Dr. Zhao Haiyuan for her support. Next, I would like to thank Professor Li Chunjin who reviewed the manuscript and made many helpful, constructive suggestions. In addition, I would like to thank my student, Miss Guo Yanlin, who checked the manuscript. Finally, I would like to express my gratitude to the editorial and production staff of Beihang University Press for their hard work in the publication of this book.

To this end, all criticism and suggestions for future improvement of the book are welcomed.

Jiang Jiahe
November, 2011

Contents

CHAPTER 1 Introduction to Missile Guidance ·· 1

   1.1 Development History of Rockets and Missiles ··· 1
   1.2 Categories of Guided Missiles ·· 2
   1.3 Missile Guidance Systems ·· 4
   1.4 Introduction to Command Guidance System ·· 6
   1.5 Introduction to Homing Guidance System ··· 8
      1.5.1 Basic Concept and Classification of Homing Guidance System ········· 8
      1.5.2 Introduction to Seeker ··· 9
   1.6 Brief Introduction to Guidance Laws ·· 13
   1.7 Autopilots ·· 14
   1.8 Outline of the Book ·· 15
   References ·· 16

CHAPTER 2 Basic Knowledge of Flight Dynamics ·· 17

   2.1 Coordinate Frames ··· 17
   2.2 Motion Parameters ··· 19
   2.3 Geometrical Parameters of Vehicle ··· 22
   2.4 Forces and Moments Acting on Vehicle ·· 23
      2.4.1 Gravity ··· 24
      2.4.2 Thrust ·· 24
      2.4.3 Aerodynamic Forces ··· 26
      2.4.4 Aerodynamic Moments ··· 30
   2.5 Hinge Moments of Control Surfaces ··· 36
   References ·· 37

CHAPTER 3 Equations of Motion for Vehicle ··· 38

   3.1 Introduction ·· 38

3.2　Dynamic Equations ·· 40
　3.2.1　Force Equations ·· 40
　3.2.2　Moment Equations ··· 41
3.3　Kinematical Equations ··· 42
　3.3.1　Kinematical Equations of the Mass Center of Vehicle ········ 42
　3.3.2　Angular Motion Equations ··································· 45
3.4　Small-disturbance Theory ··· 47
References ·· 51

CHAPTER 4　Longitudinal Motion ··· 52

4.1　State Variable Representation of the Linearized Longitudinal Equations ········ 52
4.2　Longitudinal Transfer Functions ································· 54
4.3　Longitudinal Approximations ····································· 56
　4.3.1　Short-period Approximation ·································· 56
　4.3.2　Short-period Approximation Transfer Function ············· 57
　4.3.3　Effect of Altitude and Airspeed on Short-period Mode Characteristic Parameters ·· 58
　4.3.4　Long-period Motion Approximation ························· 59
　4.3.5　Effect of Altitude and Airspeed on Long-period Mode Characteristic Parameters ·· 60
4.4　Effects of the Variation of Aerodynamic Derivatives on the Longitudinal Motion ··· 61
4.5　Solution of the Longitudinal Equations (Control Surface Locked) ········ 62
4.6　Transient Response of Vehicle ··································· 67
4.7　An Integrated Example ·· 70
References ·· 76

CHAPTER 5　Lateral Motion ·· 77

5.1　State Variable Representation of the Linearized Lateral Equations ········ 77
5.2　Lateral Transfer Functions ·· 79
5.3　Lateral Approximations ·· 83
　5.3.1　Roll Approximation ·· 83
　5.3.2　Effect of Altitude and Airspeed on Roll Mode Characteristic Parameters ··· 84

5.3.3　Dutch Roll Approximation ……………………………………………… 84
　　　5.3.4　Effect of Altitude and Airspeed on Dutch Roll Mode Characteristic
　　　　　　Parameters ………………………………………………………………… 85
　　　5.3.5　Spiral Approximation …………………………………………………… 85
　5.4　Effect of Aerodynamic Derivative Variation on Lateral Dynamics
　　　Characteristics ………………………………………………………………………… 86
　5.5　Examples of Lateral Motion ………………………………………………………… 87
　5.6　Small Disturbance Motion Equation Reduction for a Missile with Two
　　　Symmetrical Planes …………………………………………………………………… 96
　References ………………………………………………………………………………… 103

CHAPTER 6　Flight Control of Missile ……………………………………………… 104

　6.1　Introduction …………………………………………………………………………… 104
　6.2　Control Force Generation …………………………………………………………… 105
　　　6.2.1　Aerodynamic Force Control …………………………………………… 105
　　　6.2.2　Thrust Vector Control …………………………………………………… 107
　　　6.2.3　Rocket Injection Control ………………………………………………… 109
　6.3　Steering Components ………………………………………………………………… 109
　　　6.3.1　Aerodynamic Control Surfaces ………………………………………… 109
　　　6.3.2　Jet Steering Components ………………………………………………… 111
　6.4　Missile Maneuverability and Load Factor ………………………………………… 112
　6.5　Control Surface Specification ……………………………………………………… 114
　6.6　Flight Control System with Attitude Control …………………………………… 116
　　　6.6.1　Control System Components …………………………………………… 116
　　　6.6.2　Longitudinal Control ……………………………………………………… 117
　　　6.6.3　Lateral Directional Control ……………………………………………… 119
　6.7　Guidance System with Acceleration Control ……………………………………… 121
　　　6.7.1　Acceleration Control ……………………………………………………… 121
　　　6.7.2　The Two-acceleration Lateral Autopilot ……………………………… 124
　6.8　Roll Rate Stabilization ……………………………………………………………… 126
　6.9　Missile Servos ………………………………………………………………………… 126
　　　6.9.1　Pneumatic Servos ………………………………………………………… 127
　　　6.9.2　Hydraulic Servos …………………………………………………………… 129

6.9.3　Electric Servos ··· 130
6.10　Gyroscopes ··· 130
6.11　Free or Position Gyroscopes ··· 133
6.12　Rate Gyroscopes ··· 134
6.13　Accelerometers ·· 135
6.14　Altimeters ··· 137
References ··· 138

CHAPTER 7　Guidance Laws ·· 139

7.1　Motion of a Target ·· 139
7.2　Remote Control Guidance Method ······································ 142
　7.2.1　Three-point Method ·· 143
　7.2.2　Lead Angle Method ·· 149
7.3　Homing Guidance Relative Motion Equations ························ 150
7.4　Pursuit Method ··· 153
7.5　Constant-bearing Guidance ·· 155
7.6　Proportional Navigation ·· 155
References ··· 164

CHAPTER 8　Autopilot Design ··· 165

8.1　Introduction ·· 165
8.2　Autopilot of Roll Channel ··· 167
8.3　Autopilot Design Considering Body Flexibility ······················· 171
8.4　Nonlinear Stability Loop Design for Roll Channel ·················· 177
8.5　Acceleration Control System Design ···································· 182
8.6　Longitudinal Control System Design for Cruise Missile ············ 187
8.7　Lateral Control System Design for Cruise Missile ··················· 192
References ··· 198

CHAPTER 9　Command Guidance Systems ······························ 199

9.1　Principle of Command Guidance ·· 199
　9.1.1　Introduction ·· 199
　9.1.2　Actual Flight Phases ··· 203

- 9.1.3 Command Generation ... 204
- 9.2 Guidance Stations ... 207
  - 9.2.1 Basic Concepts of Radars ... 208
  - 9.2.2 Types of Guidance Stations ... 213
  - 9.2.3 Radars of Guidance Stations ... 215
  - 9.2.4 Guidance Radar Systems ... 216
- 9.3 Linear Scan Radar ... 218
  - 9.3.1 Angle Measurement ... 218
  - 9.3.2 Range-tracking Systems ... 221
  - 9.3.3 Components of Linear Scan Guidance Radar ... 223
- 9.4 Commands Transmission ... 227
  - 9.4.1 Transmission Channel ... 227
  - 9.4.2 Command Types ... 227
  - 9.4.3 Multiplex Manners of Commands ... 228
  - 9.4.4 Modulation of Commands ... 228
  - 9.4.5 Transmission Time Arrangement of Command Pulses ... 229
  - 9.4.6 Command Pulse Encoding ... 231
  - 9.4.7 Command Decoding and Demodulation ... 232
- 9.5 Brief Introduction to Optical-electronic Technique ... 233
- 9.6 Monopulse Guidance Radar ... 234
  - 9.6.1 Amplitude-comparison Monopulse ... 234
  - 9.6.2 Phase-comparison Monopulse ... 239
- 9.7 Phased-array Radar ... 241
  - 9.7.1 Principle of Phase Scanning ... 241
  - 9.7.2 Space Feed ... 245
  - 9.7.3 Phase Shifters ... 247
  - 9.7.4 Angle Measurement of Phased Array Radar ... 248
  - 9.7.5 Range and Angle Tracking System of Phased Array Radar ... 251
  - 9.7.6 Brief Introduction to Multi-function Phased Array Radar ... 252
- 9.8 Command Guidance System Design ... 253
- References ... 255

CHAPTER 10 Homing Guidance Systems ... 256

10.1 Basic Concepts of Homing Guidance ... 256
  10.1.1 Components of Homing Guidance System ... 256
  10.1.2 Classification of Homing Guidance ... 257
  10.1.3 Rate of Change of Line-of-sight ... 259
  10.1.4 Guidance Command of Homing Guidance System ... 260
10.2 Homing Heads ... 261
  10.2.1 Introduction to Homing Heads ... 261
  10.2.2 Radar Homing Heads ... 263
10.3 Semi-active Radar Homing Heads ... 264
  10.3.1 Main Technique Performance ... 264
  10.3.2 Work Principle of a Semi-active Continuous Wave Seeker ... 267
  10.3.3 Target Illumination ... 280
10.4 Brief Introduction to Active Radar Homing Head ... 280
10.5 Antenna Boresight Stabilization and Track ... 281
  10.5.1 Gyro Stabilization Platform Scheme ... 282
  10.5.2 Rate Gyro Feedback Scheme ... 283
  10.5.3 Noises Acting on Homing Heads ... 285
10.6 Infrared Seekers ... 286
  10.6.1 Infrared Radiation ... 286
  10.6.2 Introduction to Detectors ... 287
  10.6.3 Infrared Point Source Seekers ... 289
  10.6.4 Infrared Imaging Seekers ... 292
10.7 Homing Guidance System Design ... 294
  10.7.1 Homing Guidance Geometrical Relation ... 294
  10.7.2 Example of Homing Guidance System Design ... 295
10.8 Homing Guidance System Model ... 297
  10.8.1 Nonlinear Kinematical Element of Homing Guidance ... 297
  10.8.2 Block Diagram of Homing Guidance System Model ... 298
References ... 299

CHAPTER 11　Hardware-in-the-loop Simulation of Guidance and Control System ········· 300

11.1　Functions of Hardware-in-the-loop Simulation ································ 300
11.2　Hardware-in-the-loop Simulation System ·· 300
　11.2.1　Hardware-in-the-loop Simulation System Components ················ 300
　11.2.2　Subsystem Functions ·· 302
11.3　Simulation Equipments ·· 304
　11.3.1　Three-axis Flight Simulator ·· 304
　11.3.2　Hydraulic Load Simulator ·· 308
　11.3.3　Linear Acceleration Simulator ·· 310
　11.3.4　Simulation Computer ·· 311
　11.3.5　Infrared Target Simulator ·· 311

CHAPTER 1
Introduction to Missile Guidance

1.1 Development History of Rockets and Missiles

The primitive rocket was a great creation of Chinese ancient laboring people. As early as the beginning years (about A. D. 682) of Tang dynasty of China, the writings of alchemists recorded recipes about the powder and at the time the powder was invented. Until Song dynasty (about A. D. 1000) the rockets made of the powder appeared, and were used to resist invaders. This rocket consisted of an arrowhead, an arrow body, an arrow tail, and its powered device, which was a section of bamboo full of the powder and tied on the arrow body. See Figure 1.1 − 1. Although the primitive rocket is not as complicated as modern rocket, it is an embryo of modern rocket with a warhead (corresponding to an arrowhead), a propulsive system (corresponding to a bamboo full of powder), a rocket airframe (corresponding to an arrow body), and a stabilizing system (corresponding to feather). After the black powder was ignited, the arrow is shot from a bow. Obviously, the rocket could increase the range and speed of the arrow. The weapon had been used in wars until Ming dynasty about 360 years ago. In the thirteenth century, the powder, rockets, and flames invented by Chinese people were introduced into Arabia. Later, they were transferred from Arabia to Europe.

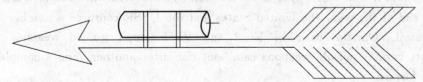

Figure 1.1 − 1 Chinese ancient rocket

After the rockets were introduced to Europe, they were developed to the rocket projectiles in the end of the eighteenth century. After the rocket techniques underwent tortuous road in the nineteenth century due to the invention of the artillery, the rocket weapon went into rapidly developing period during World War II. On October 3, 1942, Germany successfully launched V—2 missile. The V—2 was powered by a liquid-propellant rocket.

The V—2 was the first long-range, rocket-propelled missile to be put into combat. The V—2 was a supersonic missile, launched vertically and automatically tilted over a 41° to 47° angle a short time after launch. Furthermore V—2 had a liftoff weight of 12 873 kg (28 380 lb), developing a thrust of 27 125 kg (59 800 lb), a maximum acceleration 6.4g, reaching a maximum speed of about 5 705 km/h (3 545 mph), an effective range of about 354 km (220 miles), carrying a warhead of 998 kg (2 201 lb). In addition, the powered flight lasted 70 s, reaching a speed of about 1 828.8 m/s (6 000 ft/s) at burnout, with a burnout angle 45° measured from horizontal. No control was exerted after the propelling motor was shut off. Subsequently, the V—2 continued on a free-fall (ballistic) trajectory.

At that time, Germany developed aerodynamic missile V—1. The V—1, the forerunner of modern cruise missiles, was a small, midwing, pilotless monoplane, lacking ailerons but using conventional airframe and tail construction, having an overall length of 7.9 m (25.9 ft) and a wingspan of 5.3 m (17.3 ft). It weighed 2 180 kg (4 806 lb), including gasoline fuel and 850 kg (1 874 lb) warhead. Powered by a pulsejet engine and launched from inclined ramp 45.72 m (150 ft) long and 4.88 m above the ground at the highest end, the V—1 flew a preset distance, and then switched on a release system, which deflected the elevators, diving the missile straight into the ground. The engine was capable of propelling the V—1 up to 724 km/h (450 mph). A speed of 322 km/h had to be reached before the V—1 propulsion unit could maintain the missile in flight. The range of the V—1 was 370 km (230 miles). Guidance was accomplished by an autopilot along a preset path. Specifically, the plane's (or missile's) course stabilization was maintained by a magnetically controlled gyroscope that directed a tail rudder. When the predetermined distance was reached, as mentioned above, a servomechanism depressed the elevators, sending the plane into a steep dive. The V—1 was not accurate, and it was susceptible to destruction by antiaircraft fire and aircraft.

At the end of the war, the United States and the USSR captured a number of V—2s, parts and staff. On the basis of V—2 and V—1, their guided weapons got rapid development, various missile weapons came out one after another, and a complete missile system was formed.

1.2 Categories of Guided Missiles

A guided missile is defined as an unmanned vehicle which contains the warhead, is

1.2 Categories of Guided Missiles

driven by its own propulsion, depends on the guidance system to guide its flight path, and ultimately hit a target. Missile-borne engines are all jet engines which include rocket engines (solid rocket engines and liquid rocket engines), aerojet engines (turbojet engines and ramjet engines), and composite engines (solid-liquid composite engines, rocket-ramjet composite engines). Generally, a missile includes the following parts: ① a propulsion system, ② a warhead section, ③ a guidance system, and ④ an airframe. According to incomplete statistics, up to now, the types of missiles have reached more than 600 types. 200 types of them have released from military service. 400 types of them are on active service. Missiles are important weapons in modern wars. Almost all of military forces are equipped with missiles. For convenience of analysis and research, it is necessary to classify missiles. Customarily, according to the physical area of launching missile and the physical area of the target, missiles are divided into four categories: ① surface-to-surface missiles——SSM, ②surface-to-air missiles——SAM, ③ air-to-surface missiles——ASM, and ④ air-to-air missiles——AAM. We further make division as shown in Figure 1.2-1.

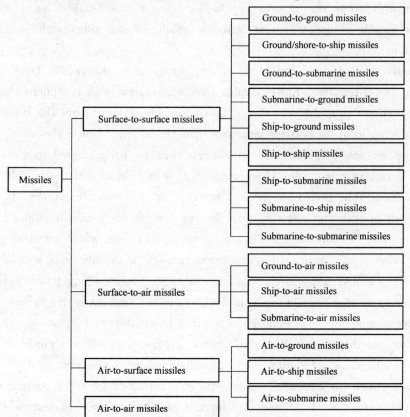

Figure 1.2-1 Categories of missiles

In addition, there are other classifications. For example, in view of fighting missions, missiles include strategic, operational, and tactical missiles. As to ranges, there are short-range missiles, medium-range missiles, intermediate-long range missiles, long-range missiles, and intercontinental missiles. In respect of warhead, missiles are divided into three categories: ①nuclear missiles, ②conventional missiles, and ③special missiles.

1.3 Missile Guidance Systems

A missile guidance system is defined as a group of components by which the missile is directed and controlled to fly toward a target. That is, a guidance system is composed of a directing subsystem and a controlling subsystem. The function of the directing subsystem is to measure the missile-target relative position, make manipulation, and then form control (i. e. guidance) commands. The function of the controlling subsystem is to execute the control commands from the directing subsystem, and hold stable flight as required. Normally the missile guidance system includes sensing, computing, stabilizing and servo-control components.

In general, a guidance process is divided into three phases: ① boost or launch, ②midcourse, and ③terminal. For example, a surface-to-air missile is launched by a booster, which is an auxiliary propulsion system. The boost phase lasts from the time the missile leaves the launcher until the booster separates from the missile. Generally, during this phase a missile may employ self-contained guidance, such as programmed guidance and INS (inertial navigation system). The characteristics of a missile vary radically before and after booster separation. The midcourse phase, when it has a distinct existence, is usually the longest in terms of both distance and time. During this phase, a missile is guided to desired course until it enters a required area (in parametric space) from which terminal guidance can successfully take over. During the midcourse phase, a missile may employ command guidance, GPS (global positioning system), or INS (inertial navigation system). The terminal phase lasts from the end of the midcourse phase to impact with a target. During this phase, command guidance or homing guidance may be employed.

According to the source of intelligence, guidance systems include self-contained guidance, non-self-contained guidance, and combined guidance systems.

In the self-contained guidance system, the generation of guidance commands does not depend on information of target and interference of external system, but rely on missile-borne instruments to measure flight data of missile and then determine flight path. For example, inertial navigation system is a typical self-contained guidance system. In inertial navigation system (INS), navigational data come from self-contained sensors (i.e., gyroscopes and accelerometers), which include a vertical accelerometer, two horizontal accelerometers, and three single-degree-of-freedom gyroscopes (or 2 two-degree-of-freedom gyroscopes). The inertial navigation system depends on integration of acceleration to obtain velocity and position. The accelerometers are mounted in gyro-stabilized inertial platform, which is used as reference. After being supplied with initial position information, the INS is capable of continuously update accurate parameters about position, ground speed, attitude, and heading. A strapdown INS employs mathematical platform instead of mechanic platform. In this way the gimbal structure is eliminated. Sensors are directly mounted on the vehicle. In strapdown INS, the transformation from the sensor to inertial reference is "computed" rather than mechanized.

In INS (including strapdown INS), since the navigation data comes from itself, the INS is insusceptible to external environment, and provides reliable all-weather navigation data. The INS is applied to ballistic and cruise missile. Other types of missiles use it for launching phase and midcourse phase.

Remote control guidance and homing guidance belong to non-self-contained guidance. The remote control guidance includes beam-riding guidance and command guidance.

Figure 1.3-1 shows an example of a beam-riding guidance system. Referring to Figure 1.3-1, there are four components: ① a tracking and guiding radar, ② a launcher, ③ a missile, and ④ a target.

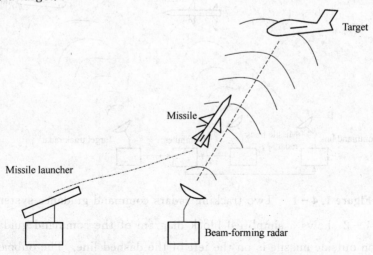

Figure 1.3-1 Beam-riding system

CHAPTER 1　Introduction to Missile Guidance

In beam rider the missile continuously rides the tracking radar beam, which aims at a target, to intercept the target. The tracking radar transmits a beam to permit a missile-borne receiver to sense both the magnitude and the direction of the error of any departure of the missile from the center of the radar beam. According to the information, the control commands are generated on board to correct flight path, and keep the missile as near as possible in the center line of the target-tracking beam till the missile destroys the target. The advantage of the beam-riding guidance technique may rely on the identical beam to guide several missiles for intercepting the same target. It requires great normal acceleration for missile due to the use of three-point guidance. Especially, if the target evasively maneuvers, the missile easily deviates from the beam. Except for radar beam, laser beam is employed.

1.4　Introduction to Command Guidance System

A command guidance system is defined as a guidance system where guidance commands come from an off-board guidance station. The guidance station is separated from the missile. Two radars track the target and the missile respectively, or a radar tracks the target and the missile. The guidance station transmits guidance commands to the missile in order to direct the missile flight at a target. Figure 1.4 – 1 indicates two tracking radars command guidance system.

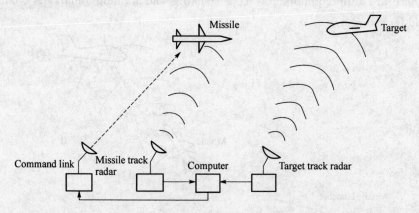

Figure 1.4 – 1　Two tracking radars command guidance system

Figure 1.4 – 2 shows a simplified block diagram of the command guidance system. A guidance station outside missile is on the left of the dashed line. The onboard equipment is on the right of the dashed line. The target tracking radar measures the elevation angle ε_T of the target, and the azimuth angle β_T of the target, whilst the missile track radar measures the elevation angle ε_M of the missile, and the azimuth angle β_M of the missile. These angles are sent to the computer. After introducing the compensator, dynamic compensation term, gravity compensation term, coordinate conversion, according to pre-selected guidance law the computer generates the elevation and azimuth control commands, k_1, k_2. After encoded, the two control commands are transmitted to the missile via a communication link. The onboard receiver receives and decodes the two signals to produce two control voltages, U_{k_1}, U_{k_2} to pitch and yaw autopilots. The autopilots control fins deflection to correct missile flight path. In view of in-flight disturbances, a missile possibly rolls. In order to guarantee the control commands to be implemented exactly, the roll position requires to be stabilized.

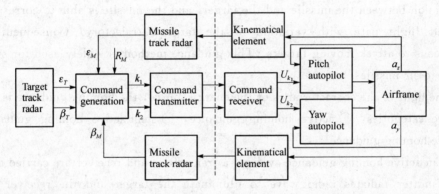

Figure 1.4 - 2 Simplified block diagram of the command guidance system

A group of radars, command generator (computer), command transmitter is termed a guidance station. The whole guidance system consists of two separate guidance loops (i. e. pitch, yaw), and a roll angle position stabilization loop. The kinematical element expresses relationship between the acceleration a_z (or a_y) and the elevation angle ε_M (or β_M). The kinematical elements form a closed loop of guidance. As long as there is angle error, $\Delta\varepsilon = \varepsilon_T - \varepsilon_M$ (or $\Delta\beta = \beta_T - \beta_M$), the guidance system tries to eliminate the error. The system is considered as a following system: the missile angle positions (ε_M, β_M) follow variation of the target angle positions (ε_T, β_T). If three-point method is employed for guidance, the range R_M from the guidance station to the missile is commonly estimated but not measured. The angle error $\Delta\varepsilon = \varepsilon_T - \varepsilon_M$ (or $\Delta\beta = \beta_T - \beta_M$) is multiplied by the range R_M so as to produce the line error $h_{\Delta\varepsilon} = \Delta\varepsilon R_M$ (or $h_{\Delta\beta} = \Delta\beta R_M$).

1.5 Introduction to Homing Guidance System

1.5.1 Basic Concept and Classification of Homing Guidance System

A homing guidance system, by means of a missile-borne seeker, directly senses the characteristic signals of target's radiation or reflection, such as infrared, radio, optical signals, and generates guidance commands to guide the missile flight at the target. There is direct relation between the missile and the target, and the missile is able to correspondingly vary itself flight path with variation of the target's trajectory. Consequently, it is advantageous to attack moving targets. This guidance method is widely used for surface-to-air or air-to-air missiles.

In the light of the characteristics of the energy source, the homing guidance is classified into three categories: ① active homing guidance, ② semi-active homing guidance, and ③ passive homing guidance.

For an active homing guidance system, a transmitter and receiver are carried on board. The transmitter radiates radio wave to illuminate the target and the receiver picks up reflection energy from the target. See Figure 1.5-1. The received energy is processed, and computed to generate guidance commands which direct the missile to intercept the target. Once the active homing head acquires a target and shifts to track the target, it is completely independent, and does not require energy transmitted from an external source. Active homing missiles have the advantage of launch-and-leave. The weight and size of missile-borne equipment increase. Because of limitation of on-board volume, an antenna aperture can not be too large, and the effective range of active seeker is not too far.

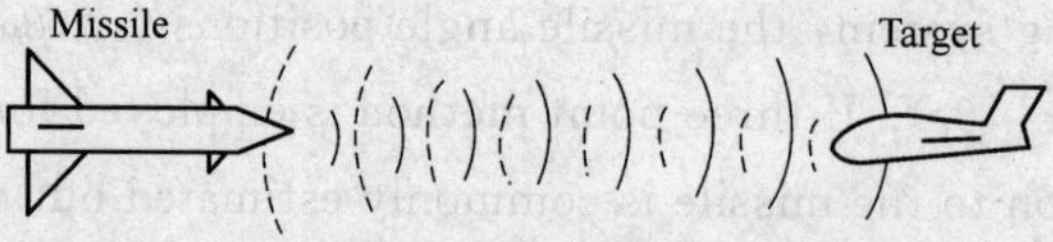

Figure 1.5-1 Active homing guidance

The difference between semi-active homing guidance and active homing guidance is that in a semi-active homing guidance system, the illuminating radar is not onboard, but on the ground, ship, or aircraft. See Figure 1.5 - 2. During guidance, semi-active homing requires the target to be continuously illuminated by the external radar at all times. The guidance system has no advantage of launch-and-leave. Because the transmitter is not onboard, but off-board, the antenna aperture may be larger. Thus the effective range of the semi-active homing guidance system is farther than that of the active homing guidance system. Since there is only one onboard receiver, compared with the active homing guidance, the weight and size of missile-borne equipment decrease.

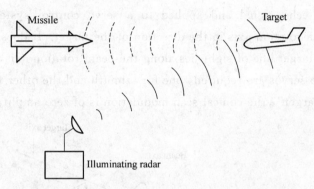

Figure 1.5 - 2 Semi-active homing guidance

A passive homing guidance directly utilizes natural emanation from a target, such as radio wave and infrared wave to guide a missile, without the need for illuminating radar. Figure 1.5 - 3 illustrates a passive homing guidance. An infrared homing guidance system is a typical example of passive homing guidance system. Passive homing missiles also have the advantage of launch-and-leave.

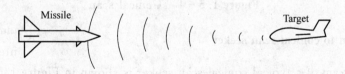

Figure 1.5 - 3 Passive homing guidance

1.5.2 Introduction to Seeker

1. Conical scan

Figure 1.5 - 4 illustrates the basic concept of conical scan. The angle between the axis of rotation and the axis of the antenna beam is the squint angle. Consider a target located at position A. Because of the rotation of the squinted beam and the target's offset from the rotation axis, the amplitude of the echo signal will be modulated at a frequency which is equal to the rotation frequency (also called the conical-scan frequency). The amplitude of the modulation depends on the angular distance between the target direction and the rotation axis. The location of the target in two angle coordinates determines the phase of the conical-scan modulation relative to the conical-scan beam rotation. The conical-scan modulation is extracted from the echo signal and applied to a servo control system that continually positions the antenna rotation axis in the direction of the target. It does this by moving the antenna so that the target line of sight lies along the beam rotation axis, as at position B in Figure 1.5 - 4. Two servos are required, one for azimuth and the other for elevation. When the antenna is "on target", the conical-scan modulation is of zero amplitude.

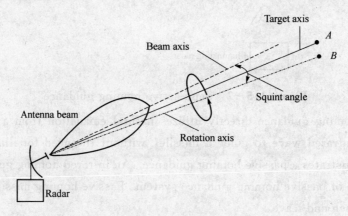

Figure 1.5 - 4 Conical scan

2. Instruction to conical-scan seeker

A block diagram of a typical conical-scan seeker is shown in Figure 1.5 - 5.

The center of the radar beam is pointed at an angle from the scan axis, and the beam is nutated so that the center of the beam follows the scan circle as it nutates. The modulation amplitude is proportional to the target offset from the rotation axis, and the phase of the modulation wave expresses deviation orientation from the rotation axis.

The reference generator is connected with the rotation motor, and provides two mutually orthogonal frequency signals. The output signal of the envelop detector is compared with the reference signals to produce elevation error and azimuth error. The two errors are sent to servos and autopilot. The error signal actuates a servosystem to drive the antenna in the proper direction to reduce the error to zero.

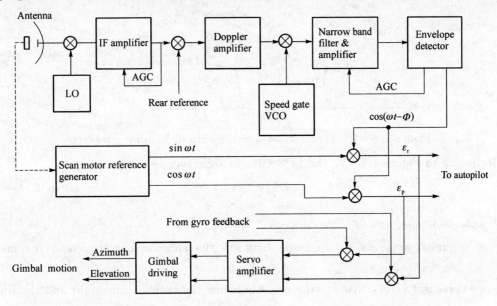

Figure 1.5 – 5   Block diagram of conical-scan seeker

3. Introduction to homing guidance system

Figure 1.5 – 6 illustrates a simplified block diagram of semi-active homing guidance system of a channel. If the missile is considered as a particle, and the effect of the radome aberration error is ignored, a simplified geometrical relationship is shown in Figure 1.5 – 7.

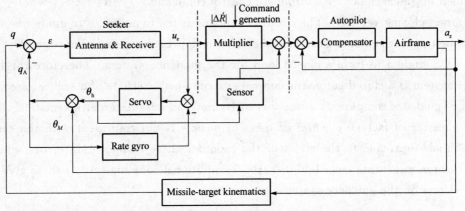

Figure 1.5 – 6   Simplified block diagram of semi-active homing guidance system

CHAPTER 1 Introduction to Missile Guidance

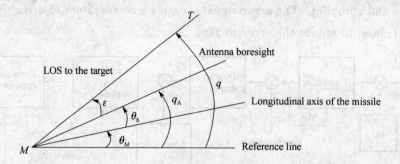

Figure 1.5-7 Geometrical relationship of homing guidance

Referring to Figure 1.5-7, the LOS (line of sight) angle q is given by
$$q=\theta_M+\theta_h+\varepsilon \tag{1.5-1}$$
where
  θ_M = pitch angle,
  θ_h = gimbal angle, i.e. an angle between the antenna boresight and the missile centerline,
  ε = boresight error, that is, the error between the antenna boresight and the line of sight to the target.

The seeker can measure the error angle ε, and give the error voltage u_ε, which is proportional to the error angle. The elevation and azimuth error voltages are used for controlling the radar beam direction. In essence, the equisignal line of the seeker follows the LOS at a rate. Thus, the error voltage u_ε is also proportional to the LOS angular rate $\dot q$. The rate gyro of the seeker is intended to stabilize the seeker boresight. In view of the requirement of guidance law, we formulate control command $k\dot q|\Delta\dot R|+k_\varphi\theta_h$, where $|\Delta\dot R|$ is missile-target closing velocity. The control command is fed to autopilot to guide the missile. The missile-target kinematics describes the relationship between the normal acceleration a_z and the LOS angle q to form a closed loop for the guidance system. Therefore the homing guidance system is a closed automatic control system which consists of a seeker, a command generator (guidance computer), autopilot, airframe, and missile-target kinematics.

As a matter of fact, the center of mass of missile is not coincident with the center of seeker. In addition, due to the effect of the radome, there is aberration error, which not only influences measuring error but also arouses additional coupling. All of these give rise to adverse effects on the guidance system.

1.6 Brief Introduction to Guidance Laws

When a missile flies toward a moving target, the missile continually corrects its flight path in accordance with a certain law, which formulates a relation between the missile and the target, or among the missile, the target and the guidance station. The law is defined as guidance law. The guidance law determines the control command to autopilot. For the identical missile and target, if different guidance laws are chosen, flight paths are different. Flight paths are expected to be as straight as possible. For example, the flight path of the lead angle method (or half lead angle method) is prior to that of the three point guidance. If the trajectory curvature is greater, then the required normal acceleration is greater, and the requirements for the aerodynamic force and structural strength and guidance system of the missile are also higher. When guidance laws are chosen, the problems of the flight time and engineering realization still need to be considered. Clearly, the flight time is expected to be as short as possible, and it is hoped that the guidance laws are easy to implement in engineering.

Guidance laws can be divided into two categories: ① classical guidance laws, and ② modern guidance laws. Although people have made great progress in research on modern guidance laws, classical guidance laws are still being used in practice. The most widely used classical guidance laws include line-of-sight guidance (i.e. three-point), lead angle method (or half lead angle method), and proportional navigation, or their improved types.

Line-of-sight guidance is also known as three-point guidance in Chinese and Russian literature. The line-of-sight guidance is one in which the missile is guided to remain on the joining the target and the control point (i.e. guidance station). Both command-line-of-sight (CLOS) and beam rider belong to the line-of-sight guidance.

The three-point guidance leads to the great trajectory curvature. In order to decrease the trajectory curvature, the lead angle method (or half lead angle method) is put forward. In essence, a compensating term is added to the three-point guidance equation.

Proportional navigation (PN) is one in which the turn rate of missile is made proportional to the turn rate of the LOS. Proportional navigation is based on the fact that if two objects are closing on each other, they will eventually collide if the LOS between the objects does not rotate in inertial space. Any rotation of the LOS (i.e., an LOS rate) is indicative of a deviation from the collision course which must be corrected by a missile maneuver. The PN guidance law seeks to null the LOS rate against nonmaneuvering targets by making the interceptor missile heading proportional to the LOS rate. Basic proportional navigation and improvement are both employed extensively in homing systems.

For highly maneuverable accelerating (or decelerating) targets, modern guidance laws, such as optimal guidance law and differential game guidance law, are superior to classical guidance laws. However, it is difficult to presently implement modern guidance laws due to requirement for more measuring parameters.

1.7 Autopilots

Generally, the function of autopilot is defined as follows: ① provide the required missile lateral acceleration response characteristics, ② stabilize or damp the bare airframe, and ③ reduce the missile performance sensitivity to disturbance inputs over the missile's flight envelope. A basic autopilot comprises a sensor, an amplifier and a servomechanism as shown in Figure 1.7 - 1. The sensors measure the missile motion parameters, such as attitude angle, angular velocity, flight altitude, acceleration and so on. Servomechanisms commonly include hydraulic, pneumatic, electrical servomechanisms. According to the magnitude and polarity of the control signal, the servomechanism deflects corresponding control surface of the missile. Autopilot and airframe form a stabilization loop.

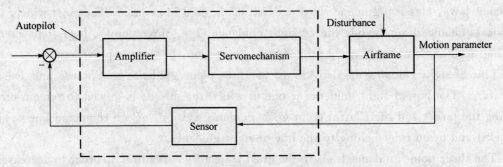

Figure 1.7 - 1 Block diagram of stabilization loop

Except for spinning missiles, autopilots generally include three channels: ① pitching, ② yawing, and ③ rolling channels. Autopilots can implement control commands: separately control pitch angle, yaw angel, roll angel, altitude, and acceleration. In addition, they can sense effects of disturbance, eliminate the effects. Autopilot, on one hand, ensures missile stability, on the other hand, ensures missile steerability.

One function of autopilot is to improve and give full play to the missile performance. For example, increase equivalent damping ratio; stabilize the missile of static instability;

decrease the influence of the dynamic parameters variation on control performances; fully utilize available normal acceleration, but ensure the acceleration of missile not to exceed the range of structural strength; and limit the angle of attack not to exceed critical angle of attack. Another function is to accurately and rapidly control a missile flight toward a target in the light of requirement. During non-guidance phase, for slantways launched missile, the mission of the autopilot is to ensure that initial dispersion to satisfy design requirement, and at the same time guarantee that the missile attitude and trajectory satisfy flight requirement of guidance phase; for vertical launched missile, the function of the autopilot is to ensure that the missile accurately and rapidly turns to the required attack direction according to programmed commands. During guidance phase, the mission of the autopilot is to accurately, rapidly, and stably control the missile flight in accordance with control commands.

Main characteristics of autopilots are as follows: ① autopilots have close relations with flight dynamics, automatic control, electronics, hydraulics, simulation technology and many other specialized technologies; and ② there are adaptability and robustness due to complexity and uncertainty of controlled objects. Autopilot design is based on classical control theory, modern control theory, or the combination of them. The most widely used design method is classical control theory. In the past, analog circuits are used for autopilot because control algorithms are simple. As control algorithms gradually become more and more complex, and computers are flexible and convenient, computers are widely used presently or in the future. The autopilot with computer is termed digital autopilot.

1.8 Outline of the Book

The following is the arrangement and contents of the book.

The book presents basic materials on tactical missile guidance. They mainly relate to three aspects: missile, control, and guidance. Firstly, controlled plant (i.e. vehicle) is researched. The forces and moments acting on a vehicle are discussed. Especially, aerodynamic forces and moments are discussed in details. After building correlative coordinate frames, based on Newton's law, the motion equations, which include dynamical equations and kinematical equations, are obtained. Based on small disturbance theory, the linear state equations of vehicle are obtained. After taking Laplace transform of the state equations, and applying Cram's rule, the transfer functions of vehicle are obtained. That is the mathematical models. On the basis of mathematical models, one will be able to design autopilot for a vehicle. Chapter 6 presents autopilot schemes. Chapter 8 gives a lot of examples of designing autopilots with MATLAB. Chapter 7 treats guidance laws, which include three-point, lead angle, pursuit, constant bearing, and proportional navigation guidance laws. Two guidance equipments, guidance radar and seeker, are described. Two guidance systems (command guidance system and homing guidance system) are treated. Chapter 9 presents command guidance equipments and systems. Chapter 10 deals with homing guidance units and systems. Chapter 11 briefly introduces guidance and control system hardware-in-the-loop simulation.

References

[1] George M Siouris. Missile Guidance and Control. New York: Spring-Verlag, 2004.
[2] Lock A S. Guidance. Princeton, New Jersey: D. Van Nostrand Company, 1955.
[3] Skolnik M I. Introduction to Radar System. 3rd ed. New York: McGraw-Hill Book Company, 2001.
[4] Skolnik M I. Radar Handbook. 2nd ed. New York: McGraw-Hill Book Company, 1990.
[5] 陈佳实. 导弹制导和控制系统的分析与设计. 北京: 宇航出版社, 1989.

CHAPTER 2
Basic Knowledge of Flight Dynamics

During flight, the velocity and attitude of vehicle will change. The variations arise from the changes of forces and moments acting on the vehicle. The forces make the vehicle translation, while the moments cause the vehicle rotation. During flight, the forces acting on vehicle include aerodynamic force, thrust of engine, and gravity. The aerodynamic force off the center of mass will create aerodynamic moment about the center of mass. The thrust not going through the center of mass will create the moment about the center of mass. Before discussion, correlative coordinate frames should be defined.

2.1　Coordinate Frames

1. Earth-surface reference frame $(S_E — O_E x_E y_E z_E)$

The frame is fixed to the earth. The origin O_E is at launch point. The $O_E x_E$ axis is horizontal and lies in nominal launch plane. The $O_E z_E$ axis points vertically downward. The $O_E y_E$ axis forms a right handed orthogonal system with the other two axes. It is shown in Figure 2.1-1.

2. Body coordinate frame $(S_b — O x_b y_b z_b)$

The frame is fixed to a vehicle. The origin O is at the mass center of the vehicle. The $O x_b$ axis points forward along longitudinal axis of the vehicle. The $O z_b$ axis is perpendicular to the $O x_b$ axis, and points downward in the plane of symmetry. The $O y_b$ axis points rightward. Three axes form a right handed orthogonal system. It is shown in Figure 2.1-1. The $O x_b$ axis is called the roll axis. The $O y_b$ axis is called the pitch axis. The $O z_b$ axis is called the yaw axis.

CHAPTER 2 Basic Knowledge of Flight Dynamics

3. Velocity coordinate system $(S_a\text{—}Ox_a y_a z_a)$

The velocity coordinate system is also called wind-axes system. The origin O is at the mass center of the vehicle. The axis Ox_a is along flight velocity. The axis Oz_a is in the longitudinal plane and is perpendicular to the Ox_a axis, and downward. The axis Oy_a forms a right handed orthogonal system with the other two axes.

The flight velocity (also called airspeed) is a moving velocity of the vehicle relative to the surrounding air.

The aerodynamic force is resolved into drag, lift and side force in this system.

4. Path coordinate frame $(S_k\text{—}Ox_k y_k z_k)$

The origin O is at the mass center of the vehicle. The axis Ox_k is along groundspeed vector. The axis Oz_k is perpendicular to the Ox_k axis, and points downward in the plumb plane including the groundspeed vector. Apparently, the Oy_k axis is in horizontal plane, and forms a right handed orthogonal system with the other two axes. The path coordinate frame is illustrated in Figure 2.1-2.

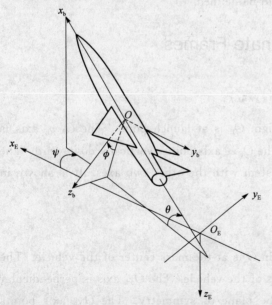

Figure 2.1-1 Body coordinate frame and Earth-surface reference frame

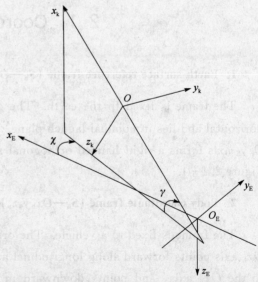

Figure 2.1-2 Velocity coordinate frame

The path velocity (or groundspeed) is a moving velocity of the vehicle relative to the ground surface.

If the groundspeed, airspeed and wind velocity are denoted by V_k, V_a and V_w respectively, then

$$V_k = V_a + V_w \qquad (2.1-1)$$

As a matter of fact, the flight velocity is not equal to the path velocity. Supposing $V_w = 0$, the flight velocity is the same as the path velocity. However, under most circumstances, we neglect the influence of wind, and consider them to be same.

2.2 Motion Parameters

1. Attitude angles or Euler angles

(1) Pitch angle θ

It is an angle between the vehicle's Ox_b axis and the $O_E x_E y_E$ plane. It is illustrated in Figure 2.1-1. If the Ox_b axis points upward relative to the $O_E x_E y_E$ plane, the angle is positive.

(2) Yaw angle ψ

It is an angle between the projection of Ox_b axis on the $O_E x_E y_E$ plane and the $O_E x_E$ axis. It is shown in Figure 2.1-1. If the projection of Ox_b axis on $O_E x_E y_E$ plane is on the right of the $O_E x_E$ axis, the angle is positive.

(3) Roll angle ϕ

It is an angle between the Oz_b axis and the plumb plane including vehicle longitudinal axis. It is illustrated in Figure 2.1-1. If the right wing banks downward about the Ox_b axis, the angle is positive.

The range of Euler angles (θ, ψ, ϕ) are limited as follows:

$$-\pi/2 \leqslant \theta \leqslant \pi/2$$
$$-\pi \leqslant \psi < \pi$$
$$-\pi \leqslant \phi < \pi$$

2. Flight path angles

(1) Elevation flight path angle γ

It is an angle between the Ox_k axis and the $O_E x_E y_E$ plane. If the Ox_k axis is upward, the

CHAPTER 2 Basic Knowledge of Flight Dynamics

angle is positive. It is illustrated in Figure 2.1-2.

(2) Azimuth flight path angle χ

It is an angle between the projection of Ox_k axis on the $O_E x_E y_E$ plane and the $O_E x_E$ axis. It is illustrated in Figure 2.1-2. If the projection of Ox_k axis on the $O_E x_E y_E$ plane is on the right of the $O_E x_E$ axis, the angle is positive.

3. The relation between velocity vector and vehicle-body coordinate system

(1) Angle of attack α

It is an angle between the projection of the velocity vector on the longitudinal symmetric plane $Ox_b z_b$ and the Ox_b axis. It is illustrated in Figure 2.2-1. If the projection of velocity vector is below the Ox_b axis, the angle of attack is positive.

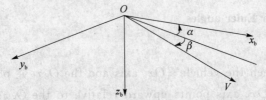

Figure 2.2-1   Angle of attack and sideslip angle

(2) Sideslip angle β

It is an angle between the velocity vector and the longitudinal symmetric plane $Ox_b z_b$. See Figure 2.2-1. If the velocity vector is at right side of the Ox_b axis, the sideslip angle is positive.

4. The relation between the path coordinate frame and the wind axes system

When the wind speed $V_w \neq 0$, the relation is complex. But when $V_w = 0$, the relation is simple. There is an angle μ (termed velocity roll angle), which is an angle between the Oz_a axis and the plumb plane including the vehicle vector. It is indicated in Figure 2.2-2. If the right wing banks downward about the Ox_b axis, the angle is positive.

5. Velocity components in the body coordinate frame

Velocity components in the body coordinate frame are expressed respectively as u, v, and w where

$u =$ component along the roll axis Ox_b,

$v =$ component along the pitch axis Oy_b,

$w =$ component along the yaw axis Oz_b.

2.2 Motion Parameters

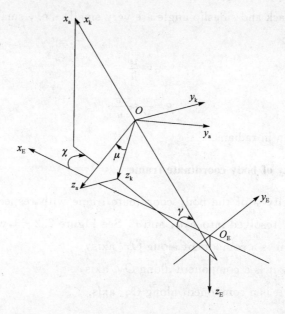

Figure 2.2-2 Relation between the path coordinate frame and the wind axes system

The components are indicated in Figure 2.2-3.

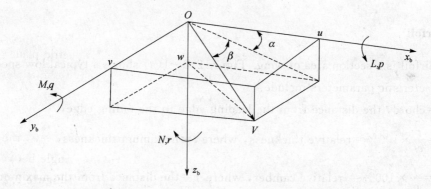

Figure 2.2-3 Velocity vector components

The Angle of attack and sideslip angle may be defined as follows:

$$\alpha = \arctan \frac{w}{u} \qquad (2.2-1)$$

$$\beta = \arcsin \frac{v}{V} \qquad (2.2-2)$$

where

$$V = (u^2 + v^2 + w^2)^{\frac{1}{2}} \qquad (2.2-3)$$

CHAPTER 2 Basic Knowledge of Flight Dynamics

If the angle of attack and sideslip angle are very small, i.e., smaller than 15°, then the α and β approximate to

$$\alpha = \frac{w}{u} \tag{2.2-4}$$

$$\beta = \frac{v}{V} \approx \frac{v}{u} \tag{2.2-5}$$

where α and β are given in radians.

6. Angular velocity of body coordinate frame

The angular velocity ω of the body coordinate frame with respect to the Earth-surface reference frame can be resolved into p, q, and r, See Figure 2.2-3, where

$p=$rate of roll, it is a component along Ox_b axis,

$q=$rate of pitch, it is a component along Oy_b axis,

$r=$rate of yaw, it is a component along Oz_b axis.

2.3 Geometrical Parameters of Vehicle

1. Airfoil

The airfoil is a section area of wing. Figure 2.3-1(a) shows a typical low speed airfoil. Main characteristic parameters include:

- $c=$chord, the distance from the leading edge to the tailing edge,
- $\bar{t} = \dfrac{t}{c} \times 100\% =$relative thickness, where $t=$maximum thickness,
- $\bar{f} = \dfrac{f}{c} \times 100\% =$ relative camber, where $f=$the distance from the maximum point of the mean camber line to the chord line.

Figure 2.3-1(b) shows supersonic airfoils. These airfoils are symmetric about the chord line. Moreover relative thickness is small.

2. Wing

Figure 2.3-2 illustrates a wing planform.

Main characteristic parameters include:

- $b=$wing span, the distance from left tip of wing to right tip of wing,

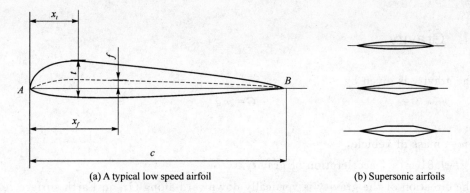

(a) A typical low speed airfoil (b) Supersonic airfoils

Figure 2.3 – 1 Typical airfoils

- S_W = wing area, wing planform area,
- $\bar{c} = \dfrac{2}{S_W}\displaystyle\int_0^{\frac{b}{2}} c^2(y)\mathrm{d}y$ = mean aerodynamic chord,
- $A = \dfrac{b^2}{S_W}$ = aspect ratio.

3. Body of missile

Main characteristic parameters of the body of missile include:

- d = diameter of missile, which is defined as the diameter of maximal sectional area of body,

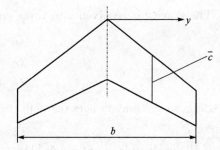

Figure 2.3 – 2 Wing planform

- $S_B = \dfrac{\pi}{4}d^2$ = maximal sectional area of body,
- l = missile body length, which is defined as distance from the nose of the missile to the tail tip of the missile.

2.4 Forces and Moments Acting on Vehicle

During flight, there are three kinds of forces acting on the vehicle. They are gravity $\boldsymbol{G}$, thrust $\boldsymbol{T}$, and resultant aerodynamic force $\boldsymbol{R}$.

When the aerodynamic force and thrusts do not pass through the center of gravity, moments are created. In the body coordinate system, the resultant moment can be divided into three components, which are called respectively the rolling moment L about Ox_b axis, the pitching moment M about Oy_b axis, and the yawing moment N about Oz_b axis. See Figure 2.2 – 3.

2.4.1 Gravity

The gravity is given by
$$G = mg$$
where
    $m =$ mass of vehicle,
    $g = 9.81$ m/s², acceleration of gravity.

The direction of the gravity is vertically downward along Oz_E in Earth surface reference frame. Thus, the components of gravity in this system are written as
$$\boldsymbol{G} = \begin{bmatrix} 0 & 0 & mg \end{bmatrix}^T$$

The gravity is resolved into three components in body coordinate frame as follows:
$$\left. \begin{aligned} G_x &= -mg\sin\theta \\ G_y &= mg\cos\theta\sin\phi \\ G_z &= mg\cos\theta\cos\phi \end{aligned} \right\} \quad (2.4-1)$$

Because the gravity acts through the center of gravity, the gravity will not generate any moments.

If flight altitude is not much high, the acceleration of gravity g is regarded as constant; if the flight altitude is much high, it will be considered that g varies with altitude.

2.4.2 Thrust

For the vehicle, the propulsion system provides thrust to push the vehicle flight. In vehicles, the aeroengines and rocket engines are used.

Some missiles such as surface-to-air missiles and anti-tank missiles are equipped with two engines. One is called the booster that is used for a missile launching and acceleration from launching carrier in initial phase, the other is called the sustainer to maintain the missile flight. Multistage rocket engines are employed for long-range missiles or intercontinental missiles.

For rocket engine, the magnitude of thrust $\boldsymbol{T}$ is given by
$$T = m_p u_e + (p_e - p_a) A_e \quad (2.4-2)$$
where
    $m_p =$ mass expelled in unit time (the propellant mass flow rate),

u_e = exhaust velocity,
p_e = exhaust pressure,
p_a = ambient pressure,
A_e = area of the exit of the engine nozzle.

Equation (2.4 - 2) shows that thrust T contains two terms: the momentum thrust and the pressure thrust. The thrust T of rocket has relationship only with flight altitude, but no relationship with other parameters of a missile. Figure 2.4 - 1 shows thrust curve of typical solid rocket engine.

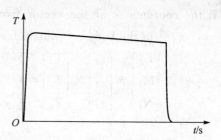

Figure 2.4 - 1 Thrust curve of a solid rocket engine

The thrust of the aerojet engine rests with flight velocity V, atmosphere density ρ, and throttle setting δ_T, so it can be expressed as

$$T = T(V, \rho, \delta_T) \qquad (2.4-3)$$

Generally, an engine is fixed in vehicle longitudinal axis direction as shown in Figure 2.4 - 2(a), sometimes in parallel direction with the longitudinal axis as shown in Figure 2.4 - 2(b). If the engine installed has any bias, or thrust vector is deflected in accordance with requirement as shown in Figure 2.4 - 2(c), then the thrust is divided into three components along three axes of body-fixed coordinate system, i.e.

$$\boldsymbol{T} = [T_x \quad T_y \quad T_z]^T \qquad (2.4-4)$$

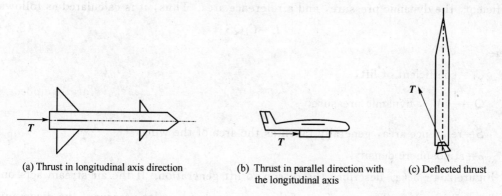

(a) Thrust in longitudinal axis direction (b) Thrust in parallel direction with the longitudinal axis (c) Deflected thrust

Figure 2.4 - 2 Acting direction of thrust T

If thrust does not pass through the center of gravity, it yields moment. The moment is resolved into

$$[L_T \quad M_T \quad N_T]^T \qquad (2.4-5)$$

If the coordinates of the vector from the mass center to the thrust acting point are $[x_T \quad y_T \quad z_T]^T$ in the body coordinate frame, then the thrust moments are expressed as

$$\begin{bmatrix} L_T \\ M_T \\ N_T \end{bmatrix} = \begin{bmatrix} x_T \\ y_T \\ z_T \end{bmatrix} \times \begin{bmatrix} T_x \\ T_y \\ T_z \end{bmatrix} = \begin{bmatrix} 0 & -z_T & y_T \\ -z_T & 0 & x_T \\ y_T & x_T & 0 \end{bmatrix} \begin{bmatrix} T_x \\ T_y \\ T_z \end{bmatrix} = \begin{bmatrix} T_z y_T - T_y z_T \\ T_x z_T - T_z x_T \\ T_y x_T - T_x y_T \end{bmatrix} \qquad (2.4-6)$$

2.4.3 Aerodynamic Forces

Whether air stream flows through a stationary object or an object moves in the air, in other words, as long as there is relative motion between an object and the air, the air produces force. The force is known as aerodynamic force. Based on the principle, a vehicle can fly in the air. Resultant aerodynamic force acting on vehicle is resolved into lift L, drag D, and side force C in wind axes system.

1. Lift L

Lift L is the component of the resultant aerodynamic force along Oz_a axis and perpendicular to the relative wind. When lift L points upward, it is positive. The aerodynamic lift is mainly produced by wings of a vehicle. The other parts of the vehicle yield a little. The aerodynamic forces are commonly expressed in terms of dimensionless coefficient, the dynamic pressure, and a reference area. Thus, it is calculated as follows:

$$L = C_L Q S \qquad (2.4-7)$$

where

C_L = coefficient of lift,

$Q = \frac{1}{2}\rho V^2$, dynamic pressure,

S = reference area, generally taken as the area of the wing,

ρ = atmosphere density.

Figure 2.4-3 (a) describes the reason to lift generation. If the air-stream acts on the airfoil at velocity V_A, the up-surface of the wing produces under-pressure, the down-surface of that produces over-pressure, and the resultant force will generate up-force, i.e., lift. If the airfoil moves in the air, certainly, the same effect will be generated. Actually, if the camber is not positive, such as a plane board, it will produce lift only if there is an angle of attack. See Figure 2.4 – 3 (b). For infinite length plane board, the coefficient of lift C_L is given by

$$C_L = 2\pi\alpha \qquad (2.4-8)$$

Generally, the lift coefficient consists of the two parts for a wing as follows:

$$C_L = C_{L_{\alpha=0}} + C_{L_\alpha}\alpha \qquad (2.4-9)$$

When calculating, we should consider the contribution of wings, fuselage, control surface, besides, coupling influences among them. Moreover, the coefficient of the lift has relationship with Mach number Ma, which is defined as ratio of the air stream speed to the local sonic speed, that is

$$Ma = \frac{V}{c} \qquad (2.4-10)$$

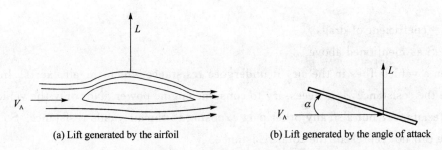

(a) Lift generated by the airfoil   (b) Lift generated by the angle of attack

Figure 2.4 – 3   Lift and air stream

When angles of attack and elevator are small, the coefficient of lift for a vehicle is expressed as

$$C_L = C_{L_0}(Ma) + C_{L_\alpha}(Ma)\alpha + C_{L_{\delta_e}}(Ma)\delta_e \qquad (2.4-11)$$

For the positive camber airfoil, C_{L_0} is positive; for the cruciform wing missile, airfoils are symmetrical, so C_{L_0} will be zero, i.e. $C_{L_0} = 0$.

Figure 2.4 – 4 indicates the C_{L_α} versus Mach number Ma curve. Ma_{cr} is critical Mach number. When flying at a lower speed ($Ma < 0.5$), C_{L_α} is considered invariable on the whole; when $0.5 < Ma < Ma_{cr}$, C_{L_α} increases in some sort; when $Ma > Ma_{cr}$, C_{L_α} increases acutely, but decreases subsequently; when $Ma > 1.5$, C_{L_α} diminishes with Ma.

2. Drag D

Drag D is the component of the resultant aerodynamic force, which is parallel to the relative wind, and points backward. Thus, it always resists a vehicle to fly forward. Analogously, drag D is written as

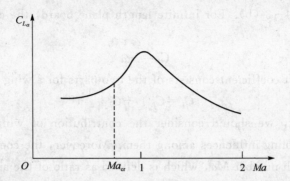

Figure 2.4-4 C_{L_α} versus Mach number Ma curve

$$D = C_D Q S \qquad (2.4-12)$$

where

C_D = coefficient of drag,

Q, S as mentioned above.

When a vehicle flies in the air, it undergoes resistance which the air exerts. In order to overcome the resistance, it is necessary to consume engine power. Not only the vehicle wings produce resistance, but also any other part exposing to the air yields resistance. So the wing resistance can not represent the total resistance.

As for a low speed vehicle, there are the friction resistance, the pressure deference resistance, and the induced resistance. For a supersonic vehicle, the shock wave results in the wave resistance. Anyway, the coefficient of drag consists of two terms as follows:

$$C_D = C_{D_0} + C_{D_i} \qquad (2.4-13)$$

where

C_{D_0} = zero lift drag coefficient,

C_{D_i} = lift induced drag coefficient.

In the case of a small angle of attack, the drag coefficient is represented as

$$C_D = C_{D_0}(Ma) + K(Ma) C_L^2 \qquad (2.4-14)$$

Moreover, the drag coefficient depends on Mach number. Figure 2.4-5 shows the C_{D_0} versus Mach number Ma curve in the case of $\alpha = 0$.

Figure 2.4-6 shows the C_L versus C_D curve, called lift-drag polar curve. It indicates that a vehicle gets lift at the cost of producing resistance.

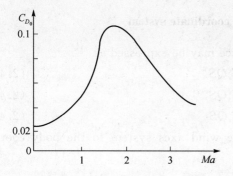

Figure 2.4-5 C_{D_0} versus Ma curve ($\alpha=0$)

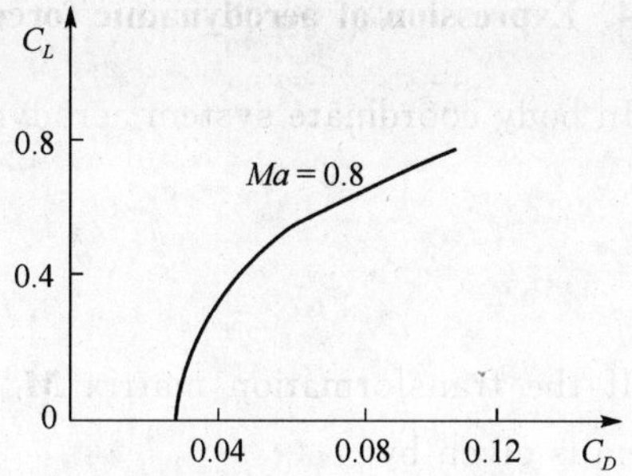

Figure 2.4-6 Lift-drag polar curve

3. Side force C

Side force C is a component of the resultant aerodynamic force in the wind-axes system. C is perpendicular to both the lift and the drag. It is calculated as follows:

$$C = C_C QS \tag{2.4-15}$$

where

C_C = coefficient of side force.

Figure 2.4-7 indicates that the vehicle flies with sideslip angle β. The side force $C(\beta)$ is expressed as

$$C(\beta) = \frac{1}{2}\rho V^2 S_W C_{Y_\beta} \beta \tag{2.4-16}$$

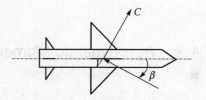

Figure 2.4-7 Sideslip angle and cross-wind force

In addition, the rudder δ_r deflection also produces side force, moreover angular rates p and r yield side force, but the latter is smaller. In conclusion, the coefficient of side force is calculated by

$$C_C = C_{C_\beta}\beta + C_{C_{\delta_r}}\delta_r + C_{C_p}\bar{p} + C_{C_r}\bar{r} \tag{2.4-17}$$

where

$C_{C_\beta} = \partial C_C / \partial \beta$,

$C_{C_{\delta_r}} = \partial C_C / \partial \delta_r$,

$C_{C_p} = \partial C_C / \partial \bar{p}, \bar{p} = \frac{pb}{2V}$,

$C_{C_r} = \partial C_C / \partial \bar{r}, \bar{r} = \frac{rb}{2V}$.

4. Expression of aerodynamic force in body coordinate system

In body coordinate system aerodynamic force may be expressed as

$$A_x = C_x QS \qquad (2.4-18)$$
$$A_y = C_y QS \qquad (2.4-19)$$
$$A_z = C_z QS \qquad (2.4-20)$$

If the transformation matrix $\boldsymbol{M}_{ba}$ from the wind axes system to the body coordinate system is given by

$$\boldsymbol{M}_{ba} = \begin{bmatrix} \cos\alpha & 0 & -\sin\alpha \\ 0 & 1 & 0 \\ \sin\alpha & 0 & \cos\alpha \end{bmatrix} \begin{bmatrix} \cos(-\beta) & \sin(-\beta) & 0 \\ -\sin(-\beta) & \cos(-\beta) & 0 \\ 0 & 0 & 1 \end{bmatrix} =$$

$$\begin{bmatrix} \cos\alpha\cos\beta & -\cos\alpha\sin\beta & -\sin\alpha \\ \sin\beta & \cos\beta & 0 \\ \sin\alpha\sin\beta & 0 & \cos\alpha \end{bmatrix} \approx \begin{bmatrix} 1 & -\beta & -\alpha \\ \beta & 1 & 0 \\ \alpha & 0 & 1 \end{bmatrix} \qquad (2.4-21)$$

then

$$\begin{bmatrix} C_x \\ C_y \\ C_z \end{bmatrix} = \boldsymbol{M}_{ba} \begin{bmatrix} -C_D \\ C_C \\ -C_L \end{bmatrix} = \begin{bmatrix} -C_D - \beta C_C + \alpha C_L \\ C_C - \beta C_D \\ -C_L - \alpha C_D \end{bmatrix} \qquad (2.4-22)$$

2.4.4 Aerodynamic Moments

1. Aerodynamic pitching moment M_A

The pitching moment is the moment about the vehicle's lateral axis (i.e., the Oy_b axis). The positive pitching moment makes the nose up. Similarly to aerodynamic forces, the pitching moment M_A can be expressed in terms of the dimensionless coefficients, flight dynamic pressure, reference area, and characteristic length as follows:

$$M_A = C_m QSl \qquad (2.4-23)$$

where

C_m = aerodynamic pitching moment coefficient,

$Q = \frac{1}{2}\rho V^2$, where ρ is atmospheric density, and V is airspeed,

S = reference area, generally taken as the vehicle's wing planform area, or sometimes maximal cross-sectional area of missile,

$l =$ characteristic length, in general taken as mean aerodynamic chord $\bar{c}$, but sometimes maximal or mean diameter of missile.

If aerodynamic configuration of a vehicle and geometrical parameters are given, the aerodynamic pitching moment depends on flight Mach number, altitude, angle of attack, and elevator deflection. In addition, if the pitch rate q, attack angle rate $\dot{\alpha}$, and elevator rate $\dot{\delta}_e$ are not zero, they create dynamic moments. Hence, the aerodynamic pitching moment is given by

$$M_A = f(V, H, \alpha, \delta_e, q, \dot{\alpha}, \dot{\delta}_e)$$

The aerodynamic pitching moment coefficient is represented as

$$C_m = C_{m_{\alpha=0}} + C_{m_\alpha}\alpha + C_{m_{\delta_e}}\delta_e + C_{m_q}\left(\frac{q\bar{c}}{2V}\right) + C_{m_{\dot{\alpha}}}\left(\frac{\dot{\alpha}\bar{c}}{2V}\right) + C_{m_{\dot{\delta}_e}}\left(\frac{\dot{\delta}_e\bar{c}}{2V}\right) \qquad (2.4-24)$$

where

$C_{m_\alpha} = \partial C_m/\partial \alpha =$ static stability derivative, i.e. pitch stiffness,

$C_{m_{\delta_e}} = \partial C_m/\partial \delta_e =$ aerodynamic pitch control derivative,

$C_{m_q} = \partial C_m/\partial[q\bar{c}/(2V)] =$ aerodynamic pitch damping derivative,

$C_{m_{\dot{\alpha}}} = \partial C_m/\partial[\dot{\alpha}\bar{c}/(2V)] =$ dynamic aerodynamic pitch derivative,

$C_{m_{\dot{\delta}_e}} = \partial C_m/\partial[\dot{\delta}_e\bar{c}/(2V)] =$ dynamic aerodynamic pitch control derivative.

These partial derivatives are nonlinear functions of Mach number.

The derivative C_{m_α} is the slope of the curve of the static pitching moment coefficient.

$C_{m_{\dot{\alpha}}}$ is the effect of the attack angle rate. The derivative results from the time lag required for the wing downwash to reach the tail. The elevator rate $\dot{\delta}_e$ corresponds to the rate of the elevator camber.

The pitch damping moment results from the pitch rate q. The damping moment is opposite from the rotating direction of the angular rate q, always opposes any pitch rate. Illustrate the pitch damping moment with the tail. If the pitch rate $q > 0$, that is, the head go up, and the tail go down, then the tail is acted by additional upward air stream. This means producing the increment of the tail attack angle. This increment generates the increment of the tail lift. The additional lift causes the increment of the pitch moment about the center of mass. The additional moment will damp the rotation of vehicle.

If letting $q = \dot{\alpha} = \dot{\delta}_e = 0$, the equation (2.4-24) becomes

$$C_m = C_{m_{\alpha=0}} + C_{m_\alpha}\alpha + C_{m_{\delta_e}}\delta_e \qquad (2.4-25)$$

From equation (2.4-25), the curve plot of C_m versus α can be obtained as shown in

Figure 2.4 - 8. In Figure 2.4 - 8, when $\alpha = \alpha_1$, the vehicle is in equilibrium. Suppose α increases i.e. $\alpha > \alpha_1$ due to a disturbance. Figure 2.4 - 8 shows the angle of attack α ($\alpha > \alpha_1$) results in the negative pitching moment, which tends to rotate the vehicle back towards the equilibrium state about the Oy_b axis. For another vehicle, suppose that the C_m versus α curve is the dashed line. Consider the case of $\alpha > \alpha_1$ due to the same disturbance. The positive pitching moment resulting from the angle of attack α ($\alpha > \alpha_1$) rotates the vehicle further away from the equilibrium state about the Oy_b axis. This simple analysis points to the conclusion: to have static longitudinal stability, the pitching moment curve must have a negative slope, i.e., $C_{m_\alpha} < 0$. C_{m_α} is also known as the pitch stiffness. In a word, if $C_{m_\alpha} < 0$, the vehicle has static longitudinal stability.

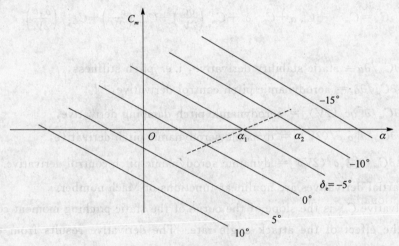

Figure 2.4 - 8 C_m versus α relation curve

The moment resulting from the elevator deflection is known as the pitch control moment. It is expressed as $C_{m_{\delta_e}} \delta_e$ in dimensionless form. In order to fly at positive angle of attack, the elevator ought to be deflected upwards (tailing edge up) for the tail control missile. For the canard missile, the elevator is deflected in opposite direction to produce the same effect.

If the airfoils are symmetrical for cruciform missiles, the $C_{m_{\alpha=0}} = 0$. In equilibrium state, the pitching moment is zero, i.e.

$$0 = C_{m_\alpha}\alpha + C_{m_{\delta_e}}\delta_e$$

or

$$\delta_{eB} = -\frac{C_{m_\alpha}}{C_{m_{\delta_e}}}\alpha_B \qquad (2.4-26)$$

The equation (2.4-26) shows that in order to maintain the equilibrium of flight state at the balancing angle of attack α_B, the deflected angle (termed balancing deflection angle, denoted by δ_{eB}) is needed. The ratio $-C_{m_\alpha}/C_{m_{\delta_e}}$ rests with aerodynamic configuration and Mach number. Experiences indicate that $-C_{m_\alpha}/C_{m_{\delta_e}}$ is taken as about 1.2 for tail control missiles, and about 1.0 for canard missiles.

2. Aerodynamic yawing moment N_A

The yawing moment N_A is the moment about the vertical axis of the vehicle (i.e. the z_b axis). A positive yawing moment will rotate the nose to the right. The aerodynamic yawing moment is generated by aerodynamic forces acting on the vehicle in lateral plane $Ox_b y_b$ which does not pass through the center of gravity. Similarly to aerodynamic pitching moment, the yawing moment N_A can be expressed in terms of the dimensionless coefficients, dynamic pressure, reference area, and characteristic length as follows:

$$N_A = C_n Q S l \qquad (2.4-27)$$

where

C_n = aerodynamic yawing moment coefficient,

$Q = \frac{1}{2}\rho V^2 =$ dynamic pressure, where $\rho =$ atmospheric density, $V =$ airspeed,

$S =$ reference area, taken as the vehicle wing planform area, or maximal cross-sectional area of missile,

$l =$ characteristic length, in general taken as the wing span b, but sometimes maximal or mean diameter of missile.

The aerodynamic yawing moment is represented as

$$N = \frac{1}{2}\rho V^2 S_W b (C_{n_\beta}\beta + C_{n_{\delta_a}}\delta_a + C_{n_{\delta_r}}\delta_r + C_{n_p}\bar{p} + C_{n_r}\bar{r}) \qquad (2.4-28)$$

where

$C_{n_\beta} = \partial C_n/\partial \beta =$ static directional stability derivative, i.e. yaw stiffness,

$C_{n_{\delta_a}} = \partial C_n/\partial \delta_a =$ aileron control coupling derivative,

$C_{n_{\delta_r}} = \partial C_n/\partial \delta_r =$ rudder control derivative,

$C_{n_p} = \partial C_n/\partial \bar{p} =$ cross derivative, where $\bar{p} = pb/2V$,

$C_{n_r} = \partial C_n/\partial \bar{r} =$ yaw damping derivative, where $\bar{r} = rb/2V$.

If $C_{n_\beta} > 0$, the vehicle has directional static stability. Suppose that the vehicle flies with a positive sideslip angle β due to a disturbance. A positive C_{n_β} can produce a positive moment, which tends to rotate the vehicle toward relative wind direction so as to decrease the sideslip angle. From the point of view, the stability is known as the weathercock stability.

The term $C_{n_p}\bar{p}$ is a yawing moment in the dimensionless coefficient due to a rolling velocity. It means there is coupling between the rolling and yawing motions.

The term $C_{n_{\delta_a}}\delta_a$ is a yawing moment in the dimensionless coefficient due to aileron deflection. For instance, if $\delta_a > 0$, the right aileron goes down, the lift increases on the right aileron, and the drag also increases simultaneously; and the left aileron goes up, the lift decreases on the left aileron, and the drag also decreases simultaneously. The drag difference of two sides produces the yawing moment.

The term $C_{n_{\delta_r}}\delta_r$ is the yaw control moment. The term $C_{n_r}\bar{r}$ is the yaw damping moment. Its effect is similar to the pitch damping moment. It always opposes any yaw rate.

These derivatives are nonlinear functions of Mach number. The equation (2.4-28) is given in terms of the missile like airplane (with a symmetric plane). For cruciform missile (with two symmetric planes), some terms may be neglected. Then the equation (2.4-28) will be reduced to

$$N = \frac{1}{2}\rho V^2 S_W b (C_{n_\beta}\beta + C_{n_{\delta_r}}\delta_r + C_{n_r}\bar{r}) \qquad (2.4-29)$$

3. Aerodynamic rolling moment L_A

The rolling moment is the moment about the longitudinal axis of the vehicle (i.e. the x_b axis). A positive rolling moment causes the right wing to move downward, and the left wing to move upward. A rolling moment is often produced by differential aerodynamic forces in sectional area, such as different lifts of two side wings, up-down different side forces, or aileron deflection.

The rolling moment L_A can be expressed in terms of the dimensionless coefficient, flight dynamic pressure, reference area, and characteristic length as follows:

$$L_A = C_l QSl \qquad (2.4-30)$$

where

C_l = aerodynamic rolling moment coefficient,

$Q = \frac{1}{2}\rho V^2$, where ρ = atmospheric density, and V = airspeed,

S = reference area, taken as vehicle's wing planform area, or maximal cross-sectional area of a missile,

l = characteristic length, in general taken as wing span b, but sometimes maximal or mean diameter of missile.

2.4 Forces and Moments Acting on Vehicle

The rolling aerodynamic moment is represented as

$$L = \frac{1}{2}\rho V^2 S_w b (C_{l_\beta}\beta + C_{l_{\delta_a}}\delta_a + C_{l_{\delta_r}}\delta_r + C_{l_p}\bar{p} + C_{l_r}\bar{r}) \qquad (2.4-31)$$

where

C_{l_β} = roll static stability derivative,

$C_{l_{\delta_a}}$ = roll control derivative,

$C_{l_{\delta_r}}$ = yaw control coupling derivative,

C_{l_p} = roll damping derivative,

C_{l_r} = cross derivative,

δ_a = aileron deflection,

δ_r = rudder deflection,

$\bar{p} = \dfrac{pb}{2V}$,

$\bar{r} = \dfrac{rb}{2V}$.

If $C_{l_\beta} < 0$, the vehicle has roll static stability derivative. In order to analyze the roll static stability, Figure 2.4-9 shows an example. If the vehicle banks a rolling angle ϕ due to a disturbance, the angle ϕ will cause lift to incline. The resultant force of the lift and gravity will cause the vehicle to sideslip in right direction, i.e., $\beta > 0$. For a positive dihedral vehicle $C_{l_\beta} < 0$ generates a negative moment to tend to rotate the vehicle back.

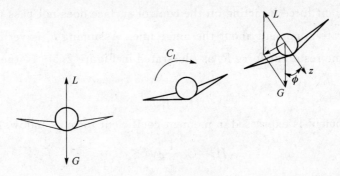

Figure 2.4-9 A vehicle automatically corrects a bank angle

The term $C_{l_{\delta_a}}\delta_a$ is the roll control moment in dimensionless form. Because the aileron deflection is differential, deflection directions of the two control surfaces are contrary, that is, the right control surface moves upward (or downward), and the left control surface moves downward (or upward).

The $C_{l_r}\bar{r}$ is the rolling moment caused by yawing angular velocity. Because $r \neq 0$, the relative airspeeds of the left and right wings are different. The lift of the wing going forward increases due to the increase of the relative airspeed, while the lift of the wing going backward decreases due to the decrease of the relative airspeed. The difference of the lifts yields the rolling moment.

The term $C_{l_{\delta_r}}\delta_r$ is the rolling moment arisen from the rudder deflection. For instance, the positive side force resulting from the rudder deflection in left direction generates a positive rolling moment because the rudder is on the body.

The term $C_{l_p}\bar{p}$ is roll damping moment. It is caused by a rolling velocity. The down-going wing increases the lift, while the up-going wing decreases the lift. The difference of lifts generates the rolling moment. It always opposes any roll rate.

2.5 Hinge Moments of Control Surfaces

If a control surface (for instance elevator, rudder, or aileron) is deflected an angle, the aerodynamic force generated on control surface produces not only a control moment about the center of gravity, but also a hinge moment about the hinge line. The hinge moment must be overcome by exerting an opposite moment, which is exerted by a pilot through a control stick for a manned vehicle, or by an actuator for an unmanned vehicle. As a matter of fact, the aerodynamic resultant force R_e acting on the control surface does not pass through the hinge line so as to generate a moment about the hinge line. Assuming h_e is vertical distance from the hinge line to the resultant force R_e as illustrated in Figure 2.5-1, the hinge moment is given by

$$H_e = R_e h_e \qquad (2.5-1)$$

The hinge moment is expressed in moment coefficient form as follows:

$$H_e = C_{h_e} \frac{1}{2} \rho V^2 S_e \bar{c}_e \qquad (2.5-2)$$

where

C_{h_e} = hinge moment coefficient,

S_e = area aft of the hinge line,

$\bar{c}_e$ = chord measured from the hinge line to trailing edge of the flap.

When the attack angle of the horizontal tail wing α_t and elevator deflection δ_e are small, for the elevator the coefficient of hinge moment is expressed as

2.5 Hinge Moments of Control Surfaces

$$C_{h_e} = \frac{\partial C_{h_e}}{\partial \alpha_t}\alpha_t + \frac{\partial C_{h_e}}{\partial \delta_e}\delta_e \qquad (2.5-3)$$

When choosing actuator, the hinge moment will have to be considered, in addition, the friction and the inertia of control mechanism.

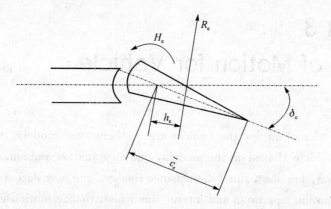

Figure 2.5 – 1 Hinge moment

References

[1] George M Siouris. Missile Guidance and Control. New York:Spring-Verlag,2004.
[2] Rafael Yanushevsky. Modern Missile Guidance. Boca Raton, Florida: Taylor & Francis Goup., 2008.
[3] John H Blakelock. Automatic Control of Aircraft and Missiles. 2nd ed. New York: John Wiley, 1991.
[4] Bern Etkin. Dynamics of Flight Stability and Control. 3rd ed. New York:John Wiley and Sons,1996.
[5] 钱杏芳,林瑞雄,赵亚男. 导弹飞行力学. 北京:北京理工大学出版社,2000.
[6] 张明廉主编. 飞行控制系统. 北京:航空工业出版社,1994.

CHAPTER 3
Equations of Motion for Vehicle

The equations of motion for the vehicle are mathematical models, which express the motion law of the vehicle. Based on the models, one may analyze and simulate the motion of a vehicle. In addition, based on small disturbance theory, one may derive linear longitudinal small disturbance motion equations and lateral small disturbance motion equations from the dynamical equations. Further, one obtains the transfer functions of a vehicle. Motions of vehicle follow Newton's laws. Newton's laws formulate the relations between the summation of external forces and the acceleration, and the relations between the summation of external moments and the angular acceleration.

3.1 Introduction

This chapter will discuss six degrees of freedom (6-DOF) nonlinear differential equations of the vehicle. These equations are obtained, based on the following assumptions:

① The earth is considered as an inertia reference, i.e., it is stationary.

② Earth's curvature is neglected, and earth-surface is assumed to be flat.

③ The vehicle is assumed to be a rigid body. Any two points on or within the airframe retain fixed with respect to each other. Ignore the aeroelastic effects of the vehicle.

④ The mass of the vehicle is assumed to retain constant.

⑤ The vehicle is considered as symmetry about $Ox_b y_b$ plane. The products of inertia I_{xy} and I_{zy} vanish.

A vector derivative in coordinate frame with an angular velocity will be discussed below. Assume that moving coordinate frame has an angular velocity $\boldsymbol{\omega}$ as shown in Figure 3.1-1. The vector $\boldsymbol{\omega}$ is resolved into three components p, q, r in this coordinate frame as follows:

$$\boldsymbol{\omega} = p\boldsymbol{i} + q\boldsymbol{j} + r\boldsymbol{k} \qquad (3.1-1)$$

where i, j, and k are unit vectors respectively along x_b, y_b, and z_b axes.

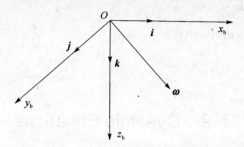

Figure 3.1-1 Components of angular velocity ω

Consider changeable vector $a(t)$. The $a(t)$ is resolved into three components a_x, a_y, a_z in the coordinate frame. Thus,

$$a = a_x i + a_y j + a_z k \qquad (3.1-2)$$

Taking derivative of $a(t)$ with respect to time t yields

$$\frac{da}{dt} = \frac{da_x}{dt}i + \frac{da_y}{dt}j + \frac{da_z}{dt}k + a_x\frac{di}{dt} + a_y\frac{dj}{dt} + a_z\frac{dk}{dt} \qquad (3.1-3)$$

Theoretical mechanics presents that if a rigid body rotates at an angular velocity ω about fixed point, the velocity of arbitrary point P in rigid body is given by

$$\frac{dr}{dt} = \omega \times r \qquad (3.1-4)$$

where r is a vector radius from original point O to the point P.

Now, the vector i is regarded as a vector radius of a point. Then

$$\frac{di}{dt} = \omega \times i \qquad (3.1-5)$$

Similarly,

$$\frac{dj}{dt} = \omega \times j \qquad (3.1-6)$$

$$\frac{dk}{dt} = \omega \times k \qquad (3.1-7)$$

Substituting the equations (3.1-5), (3.1-6), and (3.1-7) into the equation (3.1-3) yields

$$\frac{da}{dt} = \frac{da_x}{dt}i + \frac{da_y}{dt}j + \frac{da_z}{dt}k + \omega \times (a_x i + a_y j + a_z k) \qquad (3.1-8)$$

i. e.

$$\frac{da}{dt} = \frac{\delta a}{\delta t} + \omega \times a \qquad (3.1-9)$$

where

CHAPTER 3 Equations of Motion for Vehicle

$$\frac{\delta \boldsymbol{a}}{\delta t} = \frac{\mathrm{d}a_x}{\mathrm{d}t}\boldsymbol{i} + \frac{\mathrm{d}a_y}{\mathrm{d}t}\boldsymbol{j} + \frac{\mathrm{d}a_z}{\mathrm{d}t}\boldsymbol{k},$$

$\dfrac{\delta \boldsymbol{a}}{\delta t}$ is called "relative derivative",

$\dfrac{\mathrm{d}\boldsymbol{a}}{\mathrm{d}t}$ is called "absolute derivative".

3.2 Dynamic Equations

3.2.1 Force Equations

According to the Newton's second law, the dynamic equation of the vehicle is written as

$$\sum \boldsymbol{F} = \frac{\mathrm{d}}{\mathrm{d}t}(m\boldsymbol{V}) \big|_\mathrm{i} \qquad (3.2-1)$$

where

$\boldsymbol{F}$ = external force,

m = mass of vehicle,

$\boldsymbol{V}$ = mass center velocity of a vehicle,

$|_\mathrm{i}$ = denotes inertial frame.

Since the mass is assumed to be constant, the equation (3.2-1) becomes

$$\sum \boldsymbol{F} = m \frac{\mathrm{d}\boldsymbol{V}}{\mathrm{d}t} \qquad (3.2-2)$$

According to the equation (3.1-9), the equation (3.2-2) becomes

$$\sum \boldsymbol{F} = m \frac{\mathrm{d}\boldsymbol{V}}{\mathrm{d}t}\bigg|_\mathrm{B} + m(\boldsymbol{\omega} \times \boldsymbol{V}) \qquad (3.2-3)$$

where the subscript B refers to the body coordinate frame.

If $\sum \boldsymbol{F}, \boldsymbol{V}, \boldsymbol{\omega}$ are expressed respectively by components in the body coordinate frame, i.e.

$$\sum \boldsymbol{F} = F_x \boldsymbol{i} + F_y \boldsymbol{j} + F_z \boldsymbol{k} \qquad (3.2-4)$$

$$\boldsymbol{\omega} = p\boldsymbol{i} + q\boldsymbol{j} + r\boldsymbol{k} \qquad (3.2-5)$$

$$\boldsymbol{V} = u\boldsymbol{i} + v\boldsymbol{j} + w\boldsymbol{k} \qquad (3.2-6)$$

then

$$\left.\begin{array}{l}F_x = m(\dot{u}+qw-rv)\\ F_y = m(\dot{v}+ru-pw)\\ F_z = m(\dot{w}+pv-qu)\end{array}\right\} \quad (3.2-7)$$

This is the dynamic equations of mass center of a rigid vehicle in the body coordinate frame.

3.2.2 Moment Equations

From Newton's second law, the moment equation is written as

$$\sum \boldsymbol{M} = \frac{\mathrm{d}}{\mathrm{d}t}\boldsymbol{H}\Big|_i \quad (3.2-8)$$

where
- $\boldsymbol{M}=$ moment of external force,
- $\boldsymbol{H}=$ moment of momentum (angular momentum),
- $|_i=$ denotes inertial frame.

For the mass element δm, the moment of momentum is given by

$$\delta \boldsymbol{H} = \boldsymbol{r}\times(\boldsymbol{\omega}\times\boldsymbol{r})\delta m$$

Thus, the total angular momentum can be written as

$$\boldsymbol{H} = \sum \delta \boldsymbol{H} = \sum \boldsymbol{r}\times(\boldsymbol{\omega}\times\boldsymbol{r})\delta m \quad (3.2-9)$$

Let

$$\boldsymbol{r} = x\boldsymbol{i} + y\boldsymbol{j} + z\boldsymbol{k} \quad (3.2-10)$$

Substituting the equations (3.2-5) and (3.2-10) into the equation (3.2-9) yields

$$\begin{aligned}\boldsymbol{H} = &\left[p\sum(y^2+z^2)\delta m - q\sum xy\delta m - r\sum xz\delta m\right]\boldsymbol{i} + \\ &\left[-p\sum xy\delta m + q\sum(x^2+z^2)\delta m - r\sum yz\delta m\right]\boldsymbol{j} + \\ &\left[-p\sum xz\delta m - q\sum yz\delta m + r\sum(x^2+y^2)\delta m\right]\boldsymbol{k}\end{aligned} \quad (3.2-11)$$

The terms I_x, I_y, I_z are called the moments of inertia of the body about x, y, and z axes, respectively. The terms I_{xy}, I_{xz}, and I_{yz} are called the products of inertia. They are defined as follows:

$$I_x = \sum(y^2+z^2)\delta m, \quad I_{xy} = \sum xy\delta m, \quad I_y = \sum(x^2+z^2)\delta m \quad (3.2-12)$$

$$I_{xz} = \sum xz\,\mathrm{d}m, \quad I_z = \sum(x^2+y^2)\delta m, \quad I_{yz} = \sum yz\delta m \quad (3.2-13)$$

Both the moments and products of inertia depend on the shape of the body and the manner in

CHAPTER 3   Equations of Motion for Vehicle

which its mass is distributed. The larger the moments of inertia, the greater the resistance to rotation will be.

Then the scalar equations for the moment of momentum are

$$\left.\begin{array}{l} H_x = pI_x - qI_{xy} - rI_{xz} \\ H_y = -pI_{xy} + qI_y - rI_{yz} \\ H_z = -pI_{xz} - qI_{yz} + rI_z \end{array}\right\} \quad (3.2-14)$$

Using the equation (3.1-9), the equation (3.2-8) can be expanded to

$$\sum M = \left.\frac{\mathrm{d}H}{\mathrm{d}t}\right|_B + \omega \times H \quad (3.2-15)$$

Let

$$\sum M = Li + Mj + Nk \quad (3.2-16)$$

Substituting the equations (3.2-14) and (3.2-16) into the equation (3.2-15) yields

$$\left.\begin{array}{l} L = \dot{H}_x + qH_z - rH_y \\ M = \dot{H}_y + rH_x - pH_z \\ N = \dot{H}_z + pH_y - qH_x \end{array}\right\} \quad (3.2-17)$$

In general, a vehicle is symmetric about xz plane, so

$$I_{yz} = I_{xy} = 0 \quad (3.2-18)$$

Substituting the equations (3.2-14) and (3.2-18) into the equation (3.2-17) yields

$$\left.\begin{array}{l} L = I_x\dot{p} - I_{xz}\dot{r} + qr(I_z - I_y) - I_{xz}pq \\ M = I_y\dot{q} + rp(I_x - I_z) + I_{xz}(p^2 - r^2) \\ N = -I_{xz}\dot{p} + I_z\dot{r} + pq(I_y - I_x) + I_{xz}qr \end{array}\right\} \quad (3.2-19)$$

3.3   Kinematical Equations

3.3.1   Kinematical Equations of the Mass Center of Vehicle

For convenience, the origin O_E of the Earth-surface reference frame and the origin O of the body axis system are taken coincident. The translation does not change relative angles relation between the two coordinate systems. See Figure 3.3-1.

In order to discuss kinematics equations, the transformation relation between two coordinate systems are first introduced as follows:

① The transformation from the Earth-surface reference frame $S_E - O_E x_E y_E z_E$ to the transition $S'—Ox'y'z'$ through rotating a yaw angle ψ about Oz_E axis as indicated in Figure 3.3-1 is given by

$$\begin{bmatrix} x' \\ y' \\ z' \end{bmatrix} = \begin{bmatrix} \cos\psi & \sin\psi & 0 \\ -\sin\psi & \cos\psi & 0 \\ 0 & 0 & 1 \end{bmatrix} \begin{bmatrix} x_E \\ y_E \\ z_E \end{bmatrix} \quad (3.3-1)$$

② The transformation from the $S'—Ox'y'z'$ to the transition $S''—Ox''y''z''$ through rotating a pitch angle θ about Oy' axis as shown in Figure 3.3-1 is given by

$$\begin{bmatrix} x'' \\ y'' \\ z'' \end{bmatrix} = \begin{bmatrix} \cos\theta & 0 & -\sin\theta \\ 0 & 1 & 0 \\ \sin\theta & 0 & \cos\theta \end{bmatrix} \begin{bmatrix} x' \\ y' \\ z' \end{bmatrix} \quad (3.3-2)$$

③ The transformation from the $S''—Ox''y''z''$ to the body coordinate frame $S_b—Oxyz$ through rotating a roll angle ϕ about Ox'' axis as shown in Figure 3.3-1 is given by

$$\begin{bmatrix} x \\ y \\ z \end{bmatrix} = \begin{bmatrix} 1 & 0 & 0 \\ 0 & \cos\phi & \sin\phi \\ 0 & -\sin\phi & \cos\phi \end{bmatrix} \begin{bmatrix} x'' \\ y'' \\ z'' \end{bmatrix} \quad (3.3-3)$$

Therefore, the transformation from the Earth-surface coordinate frame S_E to the body coordinate frame S_b is

$$\begin{bmatrix} x \\ y \\ z \end{bmatrix} = \begin{bmatrix} 1 & 0 & 0 \\ 0 & \cos\phi & \sin\phi \\ 0 & -\sin\phi & \cos\phi \end{bmatrix} \begin{bmatrix} \cos\theta & 0 & -\sin\theta \\ 0 & 1 & 0 \\ \sin\theta & 0 & \cos\theta \end{bmatrix} \begin{bmatrix} \cos\psi & \sin\psi & 0 \\ -\sin\psi & \cos\psi & 0 \\ 0 & 0 & 1 \end{bmatrix} \begin{bmatrix} x_E \\ y_E \\ z_E \end{bmatrix} =$$

$$\begin{bmatrix} \cos\theta\cos\psi & \cos\theta\sin\psi & -\sin\theta \\ \sin\theta\cos\psi\sin\phi - \sin\psi\cos\phi & \sin\theta\sin\psi\sin\phi + \cos\psi\cos\phi & \cos\theta\sin\phi \\ \sin\theta\cos\psi\cos\phi + \sin\psi\sin\phi & \sin\theta\sin\psi\cos\phi - \cos\psi\sin\phi & \cos\theta\cos\phi \end{bmatrix} \begin{bmatrix} x_E \\ y_E \\ z_E \end{bmatrix}$$

$$(3.3-4)$$

The transformation from the body coordinate frame S_b to the Earth-surface coordinate frame S_E is given by

$$\begin{bmatrix} x_E \\ y_E \\ z_E \end{bmatrix} = \begin{bmatrix} \cos\theta\cos\psi & \cos\theta\sin\psi & -\sin\theta \\ \sin\theta\cos\psi\sin\phi - \sin\psi\cos\phi & \sin\theta\sin\psi\sin\phi + \cos\psi\cos\phi & \cos\theta\sin\phi \\ \sin\theta\cos\psi\cos\phi + \sin\psi\sin\phi & \sin\theta\sin\psi\cos\phi - \cos\psi\sin\phi & \cos\theta\cos\phi \end{bmatrix}^T \begin{bmatrix} x \\ y \\ z \end{bmatrix} =$$

$$\begin{bmatrix} \cos\theta\cos\psi & \sin\theta\cos\psi\sin\phi - \sin\psi\cos\phi & \sin\theta\cos\psi\cos\phi + \sin\psi\sin\phi \\ \cos\theta\sin\psi & \sin\theta\sin\psi\sin\phi + \cos\psi\cos\phi & \sin\theta\sin\psi\cos\phi - \cos\psi\sin\phi \\ -\sin\theta & \cos\theta\sin\phi & \cos\theta\cos\phi \end{bmatrix} \begin{bmatrix} x \\ y \\ z \end{bmatrix}$$

$$(3.3-5)$$

CHAPTER 3 Equations of Motion for Vehicle

Taking derivative of the equation (3.3-5) with respect to t yields

$$\begin{bmatrix} \dfrac{dx_E}{dt} \\ \dfrac{dy_E}{dt} \\ \dfrac{dz_E}{dt} \end{bmatrix} = \begin{bmatrix} \cos\theta\cos\psi & \sin\theta\cos\psi\sin\phi - \sin\psi\cos\phi & \sin\theta\cos\psi\cos\phi + \sin\psi\sin\phi \\ \cos\theta\sin\psi & \sin\theta\sin\psi\sin\phi + \cos\psi\cos\phi & \sin\theta\sin\psi\cos\phi - \cos\psi\sin\phi \\ -\sin\theta & \cos\theta\sin\phi & \cos\theta\cos\phi \end{bmatrix} \begin{bmatrix} \dfrac{dx}{dt} \\ \dfrac{dy}{dt} \\ \dfrac{dz}{dt} \end{bmatrix}$$

(3.3-6)

i.e.,

$$\begin{bmatrix} \dfrac{dx_E}{dt} \\ \dfrac{dy_E}{dt} \\ \dfrac{dz_E}{dt} \end{bmatrix} = \begin{bmatrix} \cos\theta\cos\psi & \sin\theta\cos\psi\sin\phi - \sin\psi\cos\phi & \sin\theta\cos\psi\cos\phi + \sin\psi\sin\phi \\ \cos\theta\sin\psi & \sin\theta\sin\psi\sin\phi + \cos\psi\cos\phi & \sin\theta\sin\psi\cos\phi - \cos\psi\sin\phi \\ -\sin\theta & \cos\theta\sin\phi & \cos\theta\cos\phi \end{bmatrix} \begin{bmatrix} u \\ v \\ w \end{bmatrix}$$

(3.3-7)

In a similar manner, the transformation from the Earth-surface coordinate frame S_E to the velocity coordinate system S_a can be written as

$$\begin{bmatrix} x_a \\ y_a \\ z_a \end{bmatrix} = \begin{bmatrix} \cos\chi\cos\gamma & \sin\chi\cos\gamma & -\sin\gamma \\ \cos\chi\sin\gamma\sin\mu - \sin\chi\cos\mu & \sin\chi\sin\gamma\sin\mu + \cos\chi\cos\mu & \cos\gamma\sin\mu \\ \cos\chi\sin\gamma\cos\mu + \sin\chi\sin\mu & \sin\chi\sin\gamma\cos\mu - \cos\chi\sin\mu & \cos\gamma\cos\mu \end{bmatrix} \begin{bmatrix} x_E \\ y_E \\ z_E \end{bmatrix}$$

(3.3-8)

Transformation from the body coordinate frame S_b to the velocity coordinate system S_a is given by

$$\begin{bmatrix} x_a \\ y_a \\ z_a \end{bmatrix} = \begin{bmatrix} \cos\alpha\cos\beta & \sin\beta & \sin\alpha\cos\beta \\ -\cos\alpha\sin\beta & \cos\beta & -\sin\alpha\sin\beta \\ -\sin\alpha & 0 & \cos\alpha \end{bmatrix} \begin{bmatrix} x \\ y \\ z \end{bmatrix}$$

(3.3-9)

Transformation relation from the velocity coordinate frame S_a to the body coordinate frame S_b is given by

$$\begin{bmatrix} x \\ y \\ z \end{bmatrix} = \begin{bmatrix} \cos\alpha\cos\beta & -\cos\alpha\sin\beta & -\sin\alpha \\ \sin\beta & \cos\beta & 0 \\ \sin\alpha\cos\beta & -\sin\alpha\sin\beta & \cos\alpha \end{bmatrix} \begin{bmatrix} x_a \\ y_a \\ z_a \end{bmatrix}$$

(3.3-10)

Then

$$\begin{bmatrix} u \\ v \\ w \end{bmatrix} = \begin{bmatrix} \cos\alpha\cos\beta & -\cos\alpha\sin\beta & -\sin\alpha \\ \sin\beta & \cos\beta & 0 \\ \sin\alpha\cos\beta & -\sin\alpha\sin\beta & \cos\alpha \end{bmatrix} \begin{bmatrix} V \\ 0 \\ 0 \end{bmatrix} = \begin{bmatrix} V\cos\alpha\cos\beta \\ V\sin\beta \\ V\sin\alpha\cos\beta \end{bmatrix}$$

(3.3-11)

Under some circumstances, the equation (3.3-11) further reduces to

$$\begin{bmatrix} u \\ v \\ w \end{bmatrix} = \begin{bmatrix} V \\ V\beta \\ V\alpha \end{bmatrix} \quad (3.3-12)$$

Substitute the equation (3.3-4) into the equation (3.3-9). The result is equal to the equation (3.3-8). Clearly, the corresponding terms in the two equations are equal, so

$$\left. \begin{array}{l} \sin \gamma = \cos \alpha \cos \beta \sin \theta - (\sin \alpha \cos \beta \cos \phi + \sin \beta \sin \phi) \cos \theta \\ \sin \chi \cos \gamma = -\cos \alpha \cos \beta \sin \psi \cos \theta + \sin \alpha \cos \beta (\cos \psi \sin \phi - \cos \phi \sin \psi \sin \theta) - \\ \qquad \sin \beta (\cos \psi \cos \phi + \sin \phi \sin \theta \sin \psi) \\ \sin \mu \cos \gamma = \cos \alpha \sin \beta \sin \theta - (\sin \alpha \sin \beta \cos \phi - \cos \beta \sin \phi) \cos \theta \end{array} \right\} \quad (3.3-13)$$

3.3.2 Angular Motion Equations

Figure 3.3-1 shows geometrical relation between angular velocity (p,q,r) and attitude rate $(\dot{\theta}, \dot{\phi}, \dot{\psi})$, where

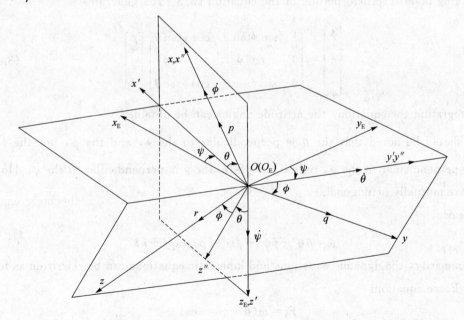

Figure 3.3-1 Geometrical relation between angular velocity (p, q, r) and attitude rate $(\dot{\theta}, \dot{\phi}, \dot{\psi})$

CHAPTER 3 Equations of Motion for Vehicle

$\dot{\psi}$ = component along Oz_E axis. If it points downwards, it is positive.

$\dot{\theta}$ = component in horizontal plane, which is perpendicular to the projection of Ox axis on the horizontal plane. If it points rightward, it is positive.

$\dot{\phi}$ = component along Ox axis. If it points forward, it is positive.

Projecting attitude rates on body axes yields

$$\left.\begin{array}{l} p = \dot{\phi} - \dot{\psi}\sin\theta \\ q = \dot{\theta}\cos\phi + \dot{\psi}\cos\theta\sin\phi \\ r = -\dot{\theta}\sin\phi + \dot{\psi}\cos\theta\cos\phi \end{array}\right\} \qquad (3.3-14)$$

i. e.

$$\begin{bmatrix} p \\ q \\ r \end{bmatrix} = \begin{bmatrix} 1 & 0 & -\sin\theta \\ 0 & \cos\phi & \cos\theta\sin\phi \\ 0 & -\sin\phi & \cos\theta\cos\phi \end{bmatrix} \begin{bmatrix} \dot{\phi} \\ \dot{\theta} \\ \dot{\psi} \end{bmatrix} \qquad (3.3-15)$$

Taking inverse transformation of the equation (3.3-15) generates

$$\begin{bmatrix} \dot{\phi} \\ \dot{\theta} \\ \dot{\psi} \end{bmatrix} = \begin{bmatrix} 1 & \sin\phi\tan\theta & \cos\phi\tan\theta \\ 0 & \cos\phi & -\sin\phi \\ 0 & \sin\phi\sec\theta & \cos\phi\sec\theta \end{bmatrix} \begin{bmatrix} p \\ q \\ r \end{bmatrix} \qquad (3.3-16)$$

Integrating the equation, the attitude angles can be obtained.

It should be noted that the $\dot{\theta}$ is perpendicular to the $\dot{\phi}$, and the $\dot{\psi}$, but the $\dot{\phi}$ is not usually perpendicular to the $\dot{\psi}$, only if the $\theta = 0$, the $\dot{\phi}$ is perpendicular to the $\dot{\psi}$. However, p, q, r are mutually orthogonal.

Clearly,

$$\boldsymbol{\omega} = \dot{\theta}\dot{\boldsymbol{\theta}}^\circ + \dot{\phi}\dot{\boldsymbol{\phi}}^\circ + \dot{\psi}\dot{\boldsymbol{\psi}}^\circ = p\boldsymbol{i} + q\boldsymbol{j} + r\boldsymbol{k} \qquad (3.3-17)$$

Summarily, the dynamic equations and kinematic equations can be rewritten as follows:

- Force equations

$$\left.\begin{array}{l} F_x = m(\dot{u} + qw - rv) \\ F_y = m(\dot{v} + ru - pw) \\ F_z = m(\dot{w} + pv - qu) \end{array}\right\} \qquad (3.3-18)$$

- Moment equations

$$\left.\begin{array}{l}L=I_x\dot{p}-I_{xz}\dot{r}+qr(I_z-I_y)-I_{xz}pq\\ M=I_y\dot{q}+rp(I_x-I_z)+I_{xz}(p^2-r^2)\\ N=-I_{xz}\dot{p}+I_z\dot{r}+pq(I_y-I_x)+I_{xz}qr\end{array}\right\} \quad (3.3-19)$$

- Angular motion equations

$$\begin{bmatrix}\dot{\phi}\\ \dot{\theta}\\ \dot{\psi}\end{bmatrix}=\begin{bmatrix}1 & \sin\phi\tan\theta & \cos\phi\tan\theta\\ 0 & \cos\phi & -\sin\phi\\ 0 & \sin\phi\sec\theta & \cos\phi\sec\theta\end{bmatrix}\begin{bmatrix}p\\ q\\ r\end{bmatrix} \quad (3.3-20)$$

- Motion equations of mass center

$$\begin{bmatrix}\dfrac{dx_E}{dt}\\ \dfrac{dy_E}{dt}\\ \dfrac{dz_E}{dt}\end{bmatrix}=\begin{bmatrix}\cos\theta\cos\psi & \sin\theta\cos\psi\sin\phi-\sin\psi\cos\phi & \sin\theta\cos\psi\cos\phi+\sin\psi\sin\phi\\ \cos\theta\sin\psi & \sin\theta\sin\psi\sin\phi+\cos\psi\cos\phi & \sin\theta\sin\psi\cos\phi-\cos\psi\sin\phi\\ -\sin\theta & \cos\theta\sin\phi & \cos\theta\cos\phi\end{bmatrix}\begin{bmatrix}u\\ v\\ w\end{bmatrix}$$

$$(3.3-21)$$

3.4 Small-disturbance Theory

The foregoing equations are nonlinear equations of motion, which completely formularize the behavior of a rigid vehicle. Although digital solutions can be obtained by digital computers, it is not convenient to directly analyze vehicle's performance based on characteristic parameters. If applying the small-disturbance theory to linearize the equations, the linearized results, which reflect the internal relations, can be obtained. Under most circumstances, the linearized results have sufficient accuracy for engineering purposes.

In applying the small-disturbance theory, it is assumed that the motion of the vehicle consists of small deviations from a reference condition of steady flight, so all the variables in the equations of motion are replaced by a reference value (denoted by subscript zero) plus a perturbation or disturbance (denoted by prefix Δ):

CHAPTER 3 Equations of Motion for Vehicle

$$\left.\begin{array}{l} u=u_0+\Delta u, v=v_0+\Delta v, w=w_0+\Delta w \\ p=p_0+\Delta p, q=q_0+\Delta q, r=r_0+\Delta r \\ X=X_0+\Delta X, Y=Y_0+\Delta Y, Z=Z_0+\Delta Z \\ M=M_0+\Delta M, N=N_0+\Delta N, L=L_0+\Delta L \\ \delta=\delta_0+\Delta\delta \\ \dot{x}_E=\dot{x}_{E0}+\Delta\dot{x}_E, \dot{y}_E=\dot{y}_{E0}+\Delta\dot{y}_E, \dot{z}_E=\dot{z}_{E0}+\Delta\dot{z}_E \\ \theta=\theta_0+\Delta\theta, \phi=\phi_0+\Delta\phi, \psi=\psi_0+\Delta\psi \end{array}\right\} \quad (3.4-1)$$

For convenience, the reference condition is assumed to be symmetric and with no angular velocity. This implies that

$$v_0=p_0=q_0=r_0=\phi_0=\psi_0=0 \quad (3.4-2)$$

Furthermore, the axis x is initially aligned with the equilibrium direction of the vehicle's velocity, and thus $w_0 = 0$. The u_0 equals the reference flight speed, and θ_0 equals the reference angle of climb (not assumed to be small). If all disturbances are set to equal zero in the foregoing equations, in the reference flight conditions, there are the following relations:

$$X_0-mg\sin\theta_0=0, \quad Y_0=0, \quad Z_0+mg\cos\theta_0=0 \quad (3.4-3)$$

$$L_0=M_0=N_0=0 \quad (3.4-4)$$

$$\dot{x}_{E0}=u_0\cos\theta_0, \quad \dot{y}_{E0}=0, \quad \dot{z}_{E0}=-u_0\sin\theta_0 \quad (3.4-5)$$

In aftermentioned discussion, the following relations of trigonometric functions will be applied:

$$\sin(\theta_0+\Delta\theta)=\sin\theta_0\cos\Delta\theta+\cos\theta_0\sin\Delta\theta=\sin\theta_0+\Delta\theta\cos\theta_0 \quad (3.4-6)$$

$$\cos(\theta_0+\Delta\theta)=\cos\theta_0\cos\Delta\theta-\sin\theta_0\sin\Delta\theta=\cos\theta_0-\Delta\theta\sin\theta_0 \quad (3.4-7)$$

In addition, the small-disturbance variables products will be neglected and only the first-order terms will be retained. As an example, the x axis force equation will be discussed as follows:

$$X-mg\sin\theta=m(\dot{u}+qw-rv) \quad (3.4-8)$$

Substituting the reference quantities and disturbance quantities into the equation (3.4-8) yields

$$X_0+\Delta X-mg\sin(\theta_0+\Delta\theta)=m\left[\frac{d}{dt}(u_0+\Delta u)+(q_0+\Delta q)(w_0+\Delta w)-(r_0+\Delta r)(v_0+\Delta v)\right] \quad (3.4-9)$$

In view of the equations (3.4-2), (3.4-3), (3.4-6), the equation (3.4-9) becomes

$$\Delta X-mg\Delta\theta\cos\theta_0=m\Delta\dot{u} \quad (3.4-10)$$

ΔX is the change in aerodynamic force and propulsive force in the Ox axis direction. The term ΔX is expanded in Taylor series in terms of the perturbation variables, u, w. That is

$$\Delta X = \frac{\partial X}{\partial u}\Delta u + \frac{\partial X}{\partial w}\Delta w + \Delta X_C \qquad (3.4-11)$$

where $\frac{\partial X}{\partial u}$ and $\frac{\partial X}{\partial w}$ are termed stability derivatives and are calculated in the reference flight conditions, and ΔX_C is the control force which results from the available controls C, such as $\Delta\delta_e$, $\Delta\delta_T$. Then,

$$\Delta X_C = \begin{bmatrix} \frac{\partial X}{\partial \delta_e} & \frac{\partial X}{\partial \delta_T} \end{bmatrix} [\Delta\delta_e \quad \Delta\delta_T]^T \qquad (3.4-12)$$

where the variable $\Delta\delta_e$ and $\Delta\delta_T$ express the changes in elevator angle and throttle setting, respectively.

Substituting the equations (3.4-11) and (3.4-12) into the equation (3.4-10) yields

$$\frac{\partial X}{\partial u}\Delta u + \frac{\partial X}{\partial w}\Delta w + \frac{\partial X}{\partial \delta_e}\Delta\delta_e + \frac{\partial X}{\partial \delta_T}\Delta\delta_T - mg\Delta\theta\cos\theta_0 = m\Delta\dot{u} \qquad (3.4-13)$$

After manipulation, the equation (3.4-13) becomes

$$\left(m\frac{\mathrm{d}}{\mathrm{d}t} - \frac{\partial X}{\partial u}\right)\Delta u - \frac{\partial X}{\partial w}\Delta w + (mg\cos\theta_0)\Delta\theta = \frac{\partial X}{\partial \delta_e}\Delta\delta_e + \frac{\partial X}{\partial \delta_T}\Delta\delta_T \qquad (3.4-14)$$

The equation (3.4-14) is divided by the mass m to become

$$\left(\frac{\mathrm{d}}{\mathrm{d}t} - X_u\right)\Delta u - X_w\Delta w + (g\cos\theta_0)\Delta\theta = X_{\delta_e}\Delta\delta_e + X_{\delta_T}\Delta\delta_T \qquad (3.4-15)$$

where

$X_u = \frac{\partial X}{\partial u}\Big/m$, $X_w = \frac{\partial X}{\partial w}\Big/m$, etc., are aerodynamic derivatives divided by the vehicle's mass.

As a matter of fact, all the motion variables influence the aerodynamic forces and moments. If only retain significant terms, then

$$\left.\begin{aligned}
\Delta X &= \frac{\partial X}{\partial u}\Delta u + \frac{\partial X}{\partial w}\Delta w + \frac{\partial X}{\partial \delta_e}\Delta\delta_e + \frac{\partial X}{\partial \delta_T}\Delta\delta_T \\
\Delta Y &= \frac{\partial Y}{\partial v}\Delta v + \frac{\partial Y}{\partial p}\Delta p + \frac{\partial Y}{\partial r}\Delta r + \frac{\partial Y}{\partial \delta_r}\Delta\delta_r \\
\Delta Z &= \frac{\partial Z}{\partial u}\Delta u + \frac{\partial Z}{\partial w}\Delta w + \frac{\partial Z}{\partial \dot{w}}\Delta\dot{w} + \frac{\partial Z}{\partial q}\Delta q + \frac{\partial Z}{\partial \delta_e}\Delta\delta_e + \frac{\partial Z}{\partial \delta_T}\Delta\delta_T
\end{aligned}\right\} \qquad (3.4-16)$$

$$\left.\begin{aligned}
\Delta L &= \frac{\partial L}{\partial v}\Delta v + \frac{\partial L}{\partial p}\Delta p + \frac{\partial L}{\partial r}\Delta r + \frac{\partial L}{\partial \delta_r}\Delta\delta_r + \frac{\partial L}{\partial \delta_a}\Delta\delta_a \\
\Delta M &= \frac{\partial M}{\partial u}\Delta u + \frac{\partial M}{\partial w}\Delta w + \frac{\partial M}{\partial \dot{w}}\Delta\dot{w} + \frac{\partial M}{\partial q}\Delta q + \frac{\partial M}{\partial \delta_e}\Delta\delta_e + \frac{\partial M}{\partial \delta_T}\Delta\delta_T \\
\Delta N &= \frac{\partial N}{\partial v}\Delta v + \frac{\partial N}{\partial p}\Delta p + \frac{\partial N}{\partial r}\Delta r + \frac{\partial N}{\partial \delta_r}\Delta\delta_r + \frac{\partial L}{\partial \delta_a}\Delta\delta_a
\end{aligned}\right\} \qquad (3.4-17)$$

CHAPTER 3 Equations of Motion for Vehicle

Similarly, other linearized small-disturbance motion equations can be obtained. The results are presented as follows.

The longitudinal small disturbance motion equations are

$$\left.\begin{aligned}\left(\frac{d}{dt}-X_u\right)\Delta u - X_w\Delta w + (g\cos\theta_0)\Delta\theta &= X_{\delta_e}\Delta\delta_e + X_{\delta_T}\Delta\delta_T \\ -Z_u\Delta u + \left[(1-Z_{\dot{w}})\frac{d}{dt}-Z_w\right]\Delta w - \left[(u_0+Z_q)\frac{d}{dt}-g\sin\theta_0\right]\Delta\theta &= Z_{\delta_e}\Delta\delta_e + Z_{\delta_T}\Delta\delta_T \\ -M_u\Delta u - \left(M_{\dot{w}}\frac{d}{dt}+M_w\right)\Delta w + \left(\frac{d^2}{dt^2}-M_q\frac{d}{dt}\right)\Delta\theta &= M_{\delta_e}\Delta\delta_e + M_{\delta_T}\Delta\delta_T\end{aligned}\right\}$$

(3.4-18)

$$\Delta\dot{x}_E = \Delta u\cos\theta_0 + \Delta w\sin\theta_0 - u_0\Delta\theta\sin\theta_0 \qquad (3.4-19)$$

$$\Delta\dot{z}_E = -\Delta u\sin\theta_0 + \Delta w\cos\theta_0 - u_0\Delta\theta\cos\theta_0 \qquad (3.4-20)$$

where

$$\Delta\dot{x}_E = \dot{x}_E - \dot{x}_{E0}$$
$$\Delta\dot{z}_E = \dot{z}_E - \dot{z}_{E0}$$

The lateral small disturbance motion equations are

$$\left.\begin{aligned}\left(\frac{d}{dt}-Y_v\right)\Delta v - Y_p\Delta p + (u_0-Y_r)\Delta r - (g\cos\theta_0)\Delta\phi &= Y_{\delta_r}\Delta\delta_r \\ -L_v\Delta v + \left(\frac{d}{dt}-L_p\right)\Delta p - \left(\frac{I_{xz}}{I_x}\frac{d}{dt}+L_r\right)\Delta r &= L_{\delta_a}\Delta\delta_a + Z_{\delta_r}\Delta\delta_r \\ -N_v\Delta v - \left(\frac{I_{xz}}{I_z}\frac{d}{dt}+N_p\right)\Delta p + \left(\frac{d}{dt}-N_r\right)\Delta r &= N_{\delta_a}\Delta\delta_a + N_{\delta_r}\Delta\delta_r\end{aligned}\right\}$$

(3.4-21)

$$\Delta\dot{\phi} = \Delta p + \Delta r\tan\theta_0 \qquad (3.4-22)$$

$$\Delta\dot{\psi} = \Delta r\sec\theta_0 \qquad (3.4-23)$$

$$\Delta\dot{y}_E = u_0\phi\cos\theta_0 + \Delta v \qquad (3.4-24)$$

where

$$\Delta\dot{y}_E = \dot{y}_E - \dot{y}_{E0}$$

Applying the small-disturbance theory, the motion equations are not only linearized, but also the equations are divided into two groups, termed longitudinal and lateral equations. This indicates that the longitudinal and lateral motions are decoupled. This facilitates problem solving, such as analysis and design of control system.

References

[1] Robert C Nelson. Flight Stability and Automatic Control. New York: McGraw-Hill, 1989.
[2] Etkin B. Dynamics of Flight Stability and Control. New York: John Wiley, 1996.
[3] John H Blakelock. Automatic Control of Aircraft and Missiles. 2nd ed. New York: John Wiley, 1991.

CHAPTER 4
Longitudinal Motion

In order to further discuss dynamic motion of a vehicle, motion models are first built. This chapter mainly discusses mathematical models about longitudinal dynamic motion to build a foundation for longitudinal motion control system design and analysis. Based on the results of chapter 3, the longitudinal disturbance motion state equations and transfer functions are derived. The disturbance motion consists of two distinct and separate modes: a long-period oscillation that is lightly damped, and a very short-period but heavily damped oscillation. Making some assumptions, approximate relationships for the long-period and short-period modes are obtained. Some examples show two motion modes of the longitudinal disturbance motion and their motion characteristics.

4.1 State Variable Representation of the Linearized Longitudinal Equations

In view of $\Delta\dot{\theta} = \Delta q$, and $\dfrac{d^2 \Delta\theta}{dt^2} = \dfrac{d}{dt}\dfrac{d\Delta\theta}{dt} = \dfrac{d\Delta q}{dt}$, the equation (3.4-18) is rewritten in the state space form as follows

$$\begin{bmatrix} \Delta\dot{u} \\ \Delta\dot{w} \\ \Delta\dot{q} \\ \Delta\dot{\theta} \end{bmatrix} = \begin{bmatrix} X_u & X_w & 0 & -g\cos\theta_0 \\ \dfrac{Z_u}{1-Z_{\dot{w}}} & \dfrac{Z_w}{1-Z_{\dot{w}}} & \dfrac{Z_q+u_0}{1-Z_{\dot{w}}} & -\dfrac{g\sin\theta_0}{1-Z_{\dot{w}}} \\ M_u + \dfrac{M_{\dot{w}} Z_u}{1-Z_{\dot{w}}} & M_w + \dfrac{M_{\dot{w}} Z_w}{1-Z_{\dot{w}}} & M_q + \dfrac{M_{\dot{w}}(Z_q+u_0)}{1-Z_{\dot{w}}} & -\dfrac{M_{\dot{w}} g\sin\theta_0}{1-Z_{\dot{w}}} \\ 0 & 0 & 1 & 0 \end{bmatrix} \begin{bmatrix} \Delta u \\ \Delta w \\ \Delta q \\ \Delta\theta \end{bmatrix} +$$

$$\begin{bmatrix} X_{\delta_e} & X_{\delta_T} \\ \dfrac{Z_{\delta_e}}{1-Z_{\dot{w}}} & \dfrac{Z_{\delta_T}}{1-Z_{\dot{w}}} \\ M_{\delta_e} + M_{\dot{w}} \dfrac{Z_{\delta_e}}{1-Z_{\dot{w}}} & M_{\delta_T} + M_{\dot{w}} \dfrac{Z_{\delta_T}}{1-Z_{\dot{w}}} \\ 0 & 0 \end{bmatrix} \begin{bmatrix} \Delta\delta_e \\ \Delta\delta_T \end{bmatrix} \qquad (4.1-1)$$

4.1 State Variable Representation of the Linearized Longitudinal Equations

Actually, Z_q and $Z_{\dot{w}}$ are small, so they are often neglected. Furthermore, assuming $\theta_0 = 0$, the equation (4.1-1) reduces to

$$\begin{bmatrix} \Delta \dot{u} \\ \Delta \dot{w} \\ \Delta \dot{q} \\ \Delta \dot{\theta} \end{bmatrix} = \begin{bmatrix} X_u & X_w & 0 & -g \\ Z_u & Z_w & u_0 & 0 \\ M_u + M_{\dot{w}} Z_u & M_w + M_{\dot{w}} Z_w & M_q + M_{\dot{w}} u_0 & 0 \\ 0 & 0 & 1 & 0 \end{bmatrix} \begin{bmatrix} \Delta u \\ \Delta w \\ \Delta q \\ \Delta \theta \end{bmatrix} +$$

$$\begin{bmatrix} X_{\delta_e} & X_{\delta_T} \\ Z_{\delta_e} & Z_{\delta_T} \\ M_{\delta_e} + M_{\dot{w}} Z_{\delta_e} & M_{\delta_T} + M_{\dot{w}} Z_{\delta_T} \\ 0 & 0 \end{bmatrix} \begin{bmatrix} \Delta \delta_e \\ \Delta \delta_T \end{bmatrix} \quad (4.1-2)$$

The force and the moment derivatives in the matrices have been divided by the mass and the moment of inertia, respectively. They are defined as:

$$X_u = \frac{\partial X/\partial u}{m}, \quad M_u = \frac{\partial M/\partial u}{I_y}, \quad \text{etc.}$$

Table 4.1-1 shows calculating expressions of the longitudinal dimensional stability derivatives.

Table 4.1-1 Longitudinal derivatives

Longitudinal derivatives	Longitudinal derivatives
$X_u = -\dfrac{(C_{D_u} + 2C_{D_0})QS}{mu_0} + \dfrac{1}{m}\dfrac{\partial T}{\partial u}$	$X_w = -\dfrac{(C_{D_\alpha} - C_{L_0})QS}{mu_0}$
$Z_u = -\dfrac{(C_{L_u} + 2C_{L_0})QS}{mu_0}$	
$Z_w = -\dfrac{(C_{L_\alpha} + C_{D_0})QS}{mu_0}$	$Z_{\dot{w}} = -C_{Z_{\dot{\alpha}}} \dfrac{\bar{c}}{2u_0} QS/(u_0 m)$
$Z_\alpha = u_0 Z_w$	$Z_{\dot{\alpha}} = u_0 Z_{\dot{w}}$
$Z_q = -C_{Z_q} \dfrac{\bar{c}}{2u_0} QS/m$	$Z_{\delta_e} = C_{Z_{\delta_e}} QS/m$
$M_u = (C_{m_u} + 2C_{m_0}) \dfrac{QS\bar{c}}{u_0 I_y}$	
$M_w = C_{m_\alpha} \dfrac{QS\bar{c}}{u_0 I_y}$	$M_{\dot{w}} = C_{m_{\dot{\alpha}}} \dfrac{\bar{c}}{2u_0} \dfrac{QS\bar{c}}{u_0 I_y}$
$M_\alpha = u_0 M_w$	$M_{\dot{\alpha}} = u_0 M_{\dot{w}}$
$M_q = C_{m_q} \dfrac{\bar{c}}{2u_0} (QS\bar{c})/I_y$	$M_{\delta_e} = C_{m_{\delta_e}} (QS\bar{c})/I_y$

CHAPTER 4 Longitudinal Motion

As an example, consider X_u. Suppose the angle between the thrust and drag is small. Then

$$X = T - D = T - C_D \frac{1}{2}\rho V^2 S$$

X_u is expressed as

$$X_u = \frac{\partial X/\partial u}{m} = -\frac{1}{m}\left(\frac{1}{2}\rho u_0^2 S \frac{\partial C_D}{\partial u} + \frac{1}{2}C_D \rho 2 u_0 S\right) + \frac{1}{m}\frac{\partial T}{\partial u} =$$

$$-\frac{1}{m}QS\left(\frac{\partial C_D}{\partial u} + 2\frac{C_{D_0}}{u_0}\right) + \frac{1}{m}\frac{\partial T}{\partial u} =$$

$$-\frac{1}{mu_0}QS\left[\frac{\partial C_D}{\partial (u/u_0)} + 2C_{D_0}\right] + \frac{1}{m}\frac{\partial T}{\partial u} = -\frac{(C_{D_u} + 2C_{D_0})QS}{mu_0} + \frac{1}{m}\frac{\partial T}{\partial u} \quad (4.1-3)$$

where the subscript 0 indicates the reference condition. The coefficient C_{D_u} can be estimated from a plot of the drag coefficient versus the Mach number curve:

$$C_{D_u} = Ma\frac{\partial C_D}{\partial Ma} \quad (4.1-4)$$

where Ma is the Mach number of interest.

4.2 Longitudinal Transfer Functions

Taking Laplace transformation of the state equation, and then using Cramer's rule, the single input single output (SISO) transfer functions can be obtained.

Taking Laplace transformation of the state equation (4.1-2) yields

$$\begin{bmatrix} s - X_u & -X_w & 0 & g \\ -Z_u & s - Z_w & -u_0 & 0 \\ -M_u - M_{\dot{w}}Z_u & -M_w - M_{\dot{w}}Z_w & s - M_q - M_{\dot{w}}u_0 & 0 \\ 0 & 0 & -1 & s \end{bmatrix} \begin{bmatrix} \Delta u \\ \Delta w \\ \Delta q \\ \Delta \theta \end{bmatrix} =$$

$$\begin{bmatrix} X_{\delta_e} & X_{\delta_T} \\ Z_{\delta_e} & Z_{\delta_T} \\ M_\delta + M_{\dot{w}}Z_\delta & M_{\delta_T} + M_{\dot{w}}Z_{\delta_T} \\ 0 & 0 \end{bmatrix} \begin{bmatrix} \Delta \delta_e \\ \Delta \delta_T \end{bmatrix} \quad (4.2-1)$$

Let $\Delta\delta_T = 0$ in the equation (4.1-1), and then apply Cramer's rule to obtain the transfer function between velocity derivation Δu and elevator deflection $\Delta\delta_e$ as follows:

4.2 Longitudinal Transfer Functions

$$\frac{\Delta u(s)}{\Delta \delta_e(s)} = \frac{\begin{vmatrix} X_{\delta_e} & -X_w & 0 & g \\ Z_{\delta_e} & s-Z_w & -u_0 & 0 \\ M_\delta + M_{\dot{w}} Z_\delta & -M_w - M_{\dot{w}} Z_w & s-M_q - M_{\dot{w}} u_0 & 0 \\ 0 & 0 & -1 & s \end{vmatrix}}{|sI - A|} =$$

$$\frac{A_u \left(s + \dfrac{1}{T_{u1}}\right)(s^2 + 2\zeta_u \omega_u s + \omega_u^2)}{(s^2 + 2\zeta_P \omega_P + \omega_P^2)(s^2 + 2\zeta_S \omega_S + \omega_S^2)} =$$

$$\frac{K_u(T_{u1}s + 1)(T_u^2 s^2 + 2\zeta_u T_u s + 1)}{(T_P^2 s^2 + 2\zeta_P T_P + 1)(T_S^2 s^2 + 2\zeta_S T_S + 1)} \qquad (4.2-2)$$

where

$A_u =$ gain of the transfer function $\Delta u/\Delta \delta_e$,

$K_u =$ transfer coefficient of the transfer function $\Delta u/\Delta \delta_e$,

$T_{u1}, T_u =$ numerator time constant,

$\zeta_u =$ numerator's damping ratio,

$\zeta_P =$ long-period or phugoid motion damping ratio,

$\zeta_S =$ short-period damping ratio,

$\omega_P =$ long-period or phugoid motion inherent frequency,

$\omega_S =$ short-period motion inherent frequency,

$T_P =$ long-period or phugoid motion time constant,

$T_S =$ short-period motion time constant.

In a similar manner, the $\Delta\theta(s)/\Delta\delta_e(s)$ can be obtained. That is

$$\frac{\Delta\theta(s)}{\Delta\delta_e(s)} = \frac{\begin{vmatrix} s-X_{\delta_e} & -X_w & 0 & X_{\delta_e} \\ -Z_u & s-Z_w & -u_0 & Z_{\delta_e} \\ -M_u - M_{\dot{w}} Z_u & -M_w - M_{\dot{w}} Z_w & s-M_q - M_{\dot{w}} u_0 & M_\delta + M_{\dot{w}} Z_\delta \\ 0 & 0 & -1 & 0 \end{vmatrix}}{|sI - A|} =$$

$$\frac{A_\theta \left(s + \dfrac{1}{T_{\theta 1}}\right)\left(s + \dfrac{1}{T_{\theta 2}}\right)}{(s^2 + 2\zeta_P \omega_P + \omega_P^2)(s^2 + 2\zeta_S \omega_S + \omega_S^2)} =$$

$$\frac{K_\theta(T_{\theta 1}s + 1)(T_{\theta 2}s + 1)}{(T_P^2 s^2 + 2\zeta_P T_P + 1)(T_S^2 s^2 + 2\zeta_S T_S + 1)} \qquad (4.2-3)$$

where

$A_\theta =$ gain of the transfer function $\Delta\theta/\Delta\delta_e$,

$K_\theta =$ transfer coefficient of the transfer function $\Delta\theta/\Delta\delta_e$,

$T_{\theta1}$, $T_{\theta2}$ = numerator time constant,

other symbols are defined ibid.

In a similar manner, the other transfer functions can be obtained.

The denominator of the transfer functions is a characteristic polynomial. In general, the characteristic polynomial consists of two terms, which respectively represent long period mode (i.e. phugoid mode) and short period mode. In a general way, the long period mode corresponds to a pair of smaller conjugate complex roots; the short period mode corresponds to a pair of greater conjugate complex roots. The typical characteristic roots distribution is presented in Figure 4.2-1.

The long period motion presents a slow flight path oscillation, and under most circumstances, under-damping. In some circumstances, the long period mode includes a positive real root and a negative real root. The positive root is unstable, and presents monotonic divergent motion. Commonly, the short period mode can not become unstable; if the center of gravity removes after aerodynamic center, the short period mode will include a positive real root and a negative real root. The positive root is unstable, and shows monotonic divergent motion; moreover the monotonic divergence index is relatively large.

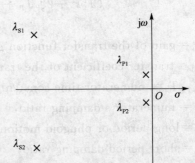

Figure 4.2-1 Typical characteristic roots distribution of longitudinal motion

4.3 Longitudinal Approximations

4.3.1 Short-period Approximation

Letting $\Delta u = 0$, and omitting X-force equation, the homogeneous longitudinal state-space equations reduces to the following:

$$\begin{bmatrix} \Delta \dot{w} \\ \Delta \dot{q} \end{bmatrix} = \begin{bmatrix} Z_w & u_0 \\ M_w + M_{\dot{w}} Z_w & M_q + M_{\dot{w}} u_0 \end{bmatrix} \begin{bmatrix} \Delta w \\ \Delta q \end{bmatrix} \qquad (4.3-1)$$

This equation can be rewritten in terms of the angle of attack using the following relations:

$$\Delta\alpha = \frac{\Delta w}{u_0}$$

$$M_\alpha = \frac{1}{I_y}\frac{\partial M}{\partial \alpha} = \frac{1}{I_y}\frac{\partial M}{\partial(\Delta w/u_0)} = \frac{u_0}{I_y}\frac{\partial M}{\partial w} = u_0 M_w$$

$$Z_\alpha = u_0 Z_w, \quad M_{\dot\alpha} = u_0 M_{\dot w}$$

Substituting the above relations into the equation (4.3-1) yields

$$\begin{bmatrix}\Delta\dot\alpha \\ \Delta\dot q\end{bmatrix} = \begin{bmatrix} \dfrac{Z_\alpha}{u_0} & 1 \\ M_\alpha + M_{\dot\alpha}\dfrac{Z_\alpha}{u_0} & M_q + M_{\dot\alpha} \end{bmatrix}\begin{bmatrix}\Delta\alpha \\ \Delta q\end{bmatrix} \qquad (4.3-2)$$

The eigenvalues of the equation (4.3-2) are the solutions of the following equation:

$$\begin{vmatrix} \lambda - \dfrac{Z_\alpha}{u_0} & -1 \\ -M_\alpha - M_{\dot\alpha}\dfrac{Z_\alpha}{u_0} & \lambda - (M_q + M_{\dot\alpha}) \end{vmatrix} = 0 \qquad (4.3-3)$$

Expanding the above determinant yields

$$\lambda^2 - \left(M_q + M_{\dot\alpha} + \frac{Z_\alpha}{u_0}\right)\lambda + M_q\frac{Z_\alpha}{u_0} - M_\alpha = 0 \qquad (4.3-4)$$

The roots of the equation (4.3-4) are given by

$$\lambda_S = \left(M_q + M_{\dot\alpha} + \frac{Z_\alpha}{u_0}\right)\bigg/2 \pm \left[\left(M_q + M_{\dot\alpha} + \frac{Z_\alpha}{u_0}\right)^2 - 4\left(M_q\frac{Z_\alpha}{u_0} - M_\alpha\right)\right]^{\frac{1}{2}}\bigg/2 \qquad (4.3-5)$$

The natural frequency and damping ratio are given by

$$\omega_S = \sqrt{M_q\frac{Z_\alpha}{u_0} - M_\alpha} \qquad (4.3-6)$$

$$\zeta_S = -\left(M_q + M_{\dot\alpha} + \frac{Z_\alpha}{u_0}\right)\bigg/(2\omega_S) \qquad (4.3-7)$$

4.3.2 Short-period Approximation Transfer Function

If the elevator acts as a control input, then the short-period approximation state space equation is

$$\begin{bmatrix}\Delta\dot\alpha \\ \Delta\dot q\end{bmatrix} = \begin{bmatrix} \dfrac{Z_\alpha}{u_0} & 1 \\ M_\alpha + M_{\dot\alpha}\dfrac{Z_\alpha}{u_0} & M_q + M_{\dot\alpha} \end{bmatrix}\begin{bmatrix}\Delta\alpha \\ \Delta q\end{bmatrix} + \begin{bmatrix} \dfrac{Z_{\delta_e}}{u_0} \\ M_{\delta_e} + \dfrac{M_{\dot\alpha} Z_{\delta_e}}{u_0} \end{bmatrix}\Delta\delta_e \qquad (4.3-8)$$

CHAPTER 4 Longitudinal Motion

Taking Laplace transformation of the equation (4.3-8) yields

$$\begin{bmatrix} s-\dfrac{Z_\alpha}{u_0} & -1 \\ -M_\alpha-M_{\dot\alpha}\dfrac{Z_\alpha}{u_0} & s-M_q-M_{\dot\alpha} \end{bmatrix} \begin{bmatrix} \Delta\alpha(s) \\ \Delta q(s) \end{bmatrix} = \begin{bmatrix} \dfrac{Z_{\delta_e}}{u_0} \\ M_{\delta_e}+\dfrac{M_{\dot\alpha} Z_{\delta_e}}{u_0} \end{bmatrix} \Delta\delta_e(s) \qquad (4.3-9)$$

The equation (4.3-9) is divided by $\Delta\delta_e(s)$ to obtain

$$\begin{bmatrix} s-\dfrac{Z_\alpha}{u_0} & -1 \\ -M_\alpha-M_{\dot\alpha}\dfrac{Z_\alpha}{u_0} & s-M_q-M_{\dot\alpha} \end{bmatrix} \begin{bmatrix} \dfrac{\Delta\alpha(s)}{\Delta\delta_e(s)} \\ \dfrac{\Delta q(s)}{\Delta\delta_e(s)} \end{bmatrix} = \begin{bmatrix} \dfrac{Z_{\delta_e}}{u_0} \\ M_{\delta_e}+\dfrac{M_{\dot\alpha} Z_{\delta_e}}{u_0} \end{bmatrix} \qquad (4.3-10)$$

Use Cramer's rule to obtain

$$\frac{\Delta\alpha(s)}{\Delta\delta_e(s)}=\frac{N^\alpha_{\delta_e}}{\Delta_{SP}}=\frac{\begin{vmatrix} \dfrac{Z_{\delta_e}}{u_0} & -1 \\ M_{\delta_e}+\dfrac{M_{\dot\alpha} Z_{\delta_e}}{u_0} & s-M_q-M_{\dot\alpha} \end{vmatrix}}{\begin{vmatrix} s-\dfrac{Z_\alpha}{u_0} & -1 \\ -M_\alpha-M_{\dot\alpha}\dfrac{Z_\alpha}{u_0} & s-M_q-M_{\dot\alpha} \end{vmatrix}}=\frac{\dfrac{Z_{\delta_e}}{u_0}s+M_{\delta_e}-M_q\dfrac{Z_{\delta_e}}{u_0}}{s^2-\left(M_q+M_{\dot\alpha}+\dfrac{Z_\alpha}{u_0}\right)s+Z_\alpha\dfrac{M_q}{u_0}-M_\alpha}$$

(4.3-11)

$$\frac{\Delta q(s)}{\Delta\delta_e(s)}=\frac{N^q_{\delta_e}}{\Delta_{SP}}=\frac{\begin{vmatrix} s-\dfrac{Z_\alpha}{u_0} & \dfrac{Z_{\delta_e}}{u_0} \\ -M_\alpha-M_{\dot\alpha}\dfrac{Z_\alpha}{u_0} & M_{\delta_e}+\dfrac{M_{\dot\alpha} Z_{\delta_e}}{u_0} \end{vmatrix}}{\begin{vmatrix} s-\dfrac{Z_\alpha}{u_0} & -1 \\ -M_\alpha-M_{\dot\alpha}\dfrac{Z_\alpha}{u_0} & s-M_q-M_{\dot\alpha} \end{vmatrix}}=\frac{\left(M_{\delta_e}+M_{\dot\alpha}\dfrac{Z_{\delta_e}}{u_0}\right)s+M_\alpha\dfrac{Z_{\delta_e}}{u_0}-M_{\delta_e}\dfrac{Z_\alpha}{u_0}}{s^2-\left(M_q+M_{\dot\alpha}+\dfrac{Z_\alpha}{u_0}\right)s+Z_\alpha\dfrac{M_q}{u_0}-M_\alpha}$$

(4.3-12)

4.3.3 Effect of Altitude and Airspeed on Short-period Mode Characteristic Parameters

The equation (4.3-6) is rewritten as

$$\omega_S = \sqrt{M_q \frac{Z_\alpha}{u_0} - M_\alpha}$$

For general vehicles,

$$M_q \frac{Z_\alpha}{u_0} \ll -M_\alpha \qquad (4.3-13)$$

Hence,

$$\omega_S = \sqrt{-M_\alpha} = \sqrt{-u_0 M_w} = \sqrt{-u_0 C_{m_\alpha} \frac{QS\bar{c}}{u_0 I_y}} = \sqrt{-C_{m_\alpha} \frac{S\bar{c}}{2I_y}} u_0 \sqrt{\rho} \qquad (4.3-14)$$

The equation (4.3-14) shows that the natural frequency of the short-period oscillation is proportional to u_0 and $\sqrt{\rho}$, where ρ decreases as the altitude increases.

In view of the equation (4.3-13), the equation (4.3-7) becomes

$$\zeta_S = -\frac{\left(M_q + M_{\dot{\alpha}} + \frac{Z_\alpha}{u_0}\right)}{2\sqrt{-M_\alpha}} \qquad (4.3-15)$$

The numerator of the expression (4.3-15) is proportional to ρu_0; the denominator of the expression is proportional to $\sqrt{\rho}\, u_0$. Thus, the damping ratio is proportional to $\sqrt{\rho}$. This means that the damping ratio decreases as the altitude increases and is approximately unchanged with the airspeed. But actually, M_α increases as Mach number increases; C_{L_α}, C_{m_q}, and $C_{m_{\dot{\alpha}}}$ increase at low Mach number phase, decrease at supersonic phase as Mach number increases. Thus, the damping eventually decreases at supersonic phase. So the damping ratio has to be enhanced by flight control system when the vehicle flies at very high altitude.

4.3.4 Long-period Motion Approximation

Compared with the short-period motion, the long-period motion responses change slowly due to its small characteristic roots. Therefore, it may be considered that the short-period dynamic behavior is over during the long-period motion. Then suppose that the change in angle of attack vanishes, i.e. $\Delta\alpha = 0$.

Due to $\Delta\alpha = \frac{\Delta w}{u_0}$, thus $\Delta w = 0$.

In addition, after neglecting the pitching moment, the homogeneous longitudinal state-space equations reduces to

$$\begin{bmatrix} \Delta\dot{u} \\ \Delta\dot{\theta} \end{bmatrix} = \begin{bmatrix} X_u & -g \\ -\dfrac{Z_u}{u_0} & 0 \end{bmatrix} \begin{bmatrix} \Delta u \\ \Delta\theta \end{bmatrix} \qquad (4.3-16)$$

CHAPTER 4 Longitudinal Motion

The eigenvalues of above equation are the solutions of the following equation:

$$\begin{vmatrix} \lambda - X_u & g \\ \dfrac{Z_u}{u_0} & \lambda \end{vmatrix} = 0 \qquad (4.3-17)$$

Expanding the above determinant yields

$$\lambda^2 - X_u \lambda - \frac{Z_u g}{u_0} = 0 \qquad (4.3-18)$$

The roots of the equation (4.3-18) are given by

$$\lambda_P = \left(X_u \pm \sqrt{X_u^2 + 4\frac{Z_u g}{u_0}} \right) \Big/ 2 \qquad (4.3-19)$$

The natural frequency and damping ratio are given by

$$\omega_P = \sqrt{-\frac{Z_u g}{u_0}} \qquad (4.3-20)$$

$$\zeta_P = -\frac{X_u}{2\omega_P} \qquad (4.3-21)$$

4.3.5 Effect of Altitude and Airspeed on Long-period Mode Characteristic Parameters

From Table 4.1-1,

$$Z_u = -\frac{(C_{L_u} + 2C_{L_0})QS}{mu_0} \qquad (4.3-22)$$

For low Mach number,

$$C_{L_u} = \frac{\partial C_L}{\partial u/u_0} = \frac{u_0}{c}\frac{\partial C_L}{\partial u/c} = Ma\,\frac{\partial C_L}{\partial Ma} = 0 \qquad (4.3-23)$$

If the vehicle is in equilibrium state, then the lift is equal to the gravity, i.e.

$$C_L QS = mg$$

Then

$$\omega_P = \sqrt{-\frac{Z_u g}{u_0}} = \sqrt{\frac{2C_{L_0} QSg}{mu_0 u_0}} = \sqrt{\frac{2mgg}{mu_0^2}} = \frac{\sqrt{2}g}{u_0} \qquad (4.3-24)$$

and

$$\zeta_P = -\frac{X_u}{2\omega_P} = -\frac{-(C_{D_u} + 2C_{D_0})QS/mu_0}{2\sqrt{2}\,\dfrac{g}{u_0}} =$$

$$\frac{(0+2C_{D_0})QS/mu_0}{2\sqrt{2}\,\frac{g}{u_0}} = \frac{C_{D_0}QS/mu_0}{\sqrt{2}mg/mu_0} = \frac{D}{\sqrt{2}G} = \frac{D}{\sqrt{2}L} = \frac{C_D}{\sqrt{2}C_L} \qquad (4.3-25)$$

An examination of the equation (4.3-24) indicates that the natural frequency of the long-period mode is inversely proportional to the forward speed and is approximately independent of ρ.

An observation of the equation (4.3-25) shows that the damping ratio is inversely proportional to the lift-to-drag ratio. It can be seen from the equation (4.3-25) that the phugoid damping is degraded as the aerodynamic efficiency (L/D) is increased. To improve the damping of the phugoid motion, the design would have to reduce the lift-to-drag ratio of the vehicle. Because this would degrade the performance of the vehicle, the designer would find such a choice unacceptable and would look for another alternative, such as an automatic stabilization system to provide the proper damping characteristics. If the vehicle is in straight and level flight, the lift equals the gravity i.e. constant. Under the circumstance the damping ratio of the long-period is proportional to the total drag. When the airspeed increases at a given altitude, the drag increases and thus the long-period damping ratio should increase. Similarly, if the airspeed retains constant but the altitude is increased, the long-period damping ratio should decrease due to the decrease in drag resulting from the lower atmosphere density at the higher altitude.

The only usefulness of the long-period approximation is to approximately evaluate the natural frequency and the damping ratio of the long-period mode. The approximate transfer function of long-period motion can not accurately reproduce the long-period oscillation. The magnitudes of the oscillations are much more smaller than those obtained from the complete longitudinal equations; furthermore, there is 180° phase shift between them. The long-period approximation is not satisfactory for simulation purposes.

4.4 Effects of the Variation of Aerodynamic Derivatives on the Longitudinal Motion

The equation (4.3-14) shows that if increasing $|C_{m_a}|$, the short-period natural frequency ω_S increases. The equation (4.3-15) indicates that if increasing $|C_{L_a}|$, $|C_{m_q}|$, or $|C_{m_{\dot{a}}}|$, the damping ratio of short-period increases, and if increasing $|C_{m_a}|$, the damping ratio of the short-period decreases.

CHAPTER 4 Longitudinal Motion

The equation (4.3 – 20) indicates that if increasing $|C_{Z_u}|$, the long-period natural frequency increases. The equation (4.3 – 25) indicates that if increasing $|C_{X_u}|$, the long-period damping ratio increases. Table 4.4 – 1 summarizes the effect of each derivative on the longitudinal motion.

Table 4.4 – 1 Influence of aerodynamic derivatives on the long and short period motions

Aerodynamic derivatives	Quantity most affected	How affected
$\|C_{m_q}\|$	Damping of short-period, ζ_S	Increase $\|C_{m_q}\|$ to increase the damping
$\|C_{m_{\dot{\alpha}}}\|$	Damping of short-period, ζ_S	Increase $\|C_{m_{\dot{\alpha}}}\|$ to increase the damping
$\|C_{L_\alpha}\|$	Damping of short-period, ζ_S	Increase $\|C_{L_\alpha}\|$ to increase the damping
$\|C_{m_\alpha}\|$	Natural frequency of the short-period, ω_S	Increase $\|C_{m_\alpha}\|$ to increase the frequency
$\|C_{Z_u}\|$	Natural frequency of the long-period, ω_P	Increase $\|C_{Z_u}\|$ to increase the frequency
$\|C_{X_u}\|$	Damping of the long-period, ζ_P	Increase $\|C_{X_u}\|$ to increase the damping

4.5 Solution of the Longitudinal Equations (Control Surface Locked)

Here, only consider the homogeneous equations. This means no external inputs: $\delta_e = \delta_T = 0$. Then the equation (4.1 – 1) becomes

$$\begin{bmatrix} \Delta \dot{u} \\ \Delta \dot{w} \\ \Delta \dot{q} \\ \Delta \dot{\theta} \end{bmatrix} = \begin{bmatrix} X_u & X_w & 0 & -g\cos\theta_0 \\ \dfrac{Z_u}{1-Z_{\dot{w}}} & \dfrac{Z_w}{1-Z_{\dot{w}}} & \dfrac{Z_q+u_0}{1-Z_{\dot{w}}} & -\dfrac{g\sin\theta_0}{1-Z_{\dot{w}}} \\ M_u+\dfrac{M_{\dot{w}}Z_u}{1-Z_{\dot{w}}} & M_w+\dfrac{M_{\dot{w}}Z_w}{1-Z_{\dot{w}}} & M_q+\dfrac{M_{\dot{w}}(Z_q+u_0)}{1-Z_{\dot{w}}} & -\dfrac{M_{\dot{w}}g\sin\theta_0}{1-Z_{\dot{w}}} \\ 0 & 0 & 1 & 0 \end{bmatrix} \begin{bmatrix} \Delta u \\ \Delta w \\ \Delta q \\ \Delta \theta \end{bmatrix} \quad (4.5-1)$$

Example 4.5 – 1: A surface-to-air missile is flying at altitude $H = 5\,000$ m with a flight velocity $V = 641$ m/s. The homogeneous equation is given by

$$\begin{bmatrix} \Delta \dot{u} \\ \Delta \dot{q} \\ \Delta \dot{\alpha} \\ \Delta \dot{\theta} \end{bmatrix} = \begin{bmatrix} -0.003\,98 & 0 & -24.32 & -7.73 \\ 0.000\,009\,351 & -1.163\,3 & -102.021\,955\,8 & -0.001\,442\,553 \\ -0.000\,061\,5 & 1 & -1.161\,41 & 0.009\,41 \\ 0 & 1 & 0 & 0 \end{bmatrix} \begin{bmatrix} \Delta u \\ \Delta q \\ \Delta \alpha \\ \Delta \theta \end{bmatrix}$$

Use MATLAB to solve this example. MATLAB program 4.5 – 1 generates the coefficients of the characteristic polynomial and eigenvalues of the coefficient matrix.

4.5 Solution of the Longitudinal Equations (Control Surface Locked)

MATLAB program 4.5 - 1

```
>>a=[-0.00398 0 -24.32 -7.73;
     0.000009351 -1.1633 -102.0219558 -0.001442553;
     -0.0000615 1 -1.16141 0.00941;
     0 1 0 0];
>> poly(a)
ans =
    1.0000    2.3287   103.3822    1.3717    0.0524
>> eig(a)
ans =
    -1.1577+10.1000i
    -1.1577-10.1000i
    -0.0066+0.0215i
    -0.0066-0.0215i
```

Then the characteristic equation $|\lambda I - A| = 0$ is directly written as
$$\lambda^4 + 2.3287\lambda^3 + 103.3822\lambda^2 + 1.3717\lambda + 0.0524 = 0$$
The eigenvalues of the characteristic equation are as follows:
$$\lambda_{1,2} = -0.0066 \pm i(0.0215) \quad \text{(long-period)}$$
$$\lambda_{3,4} = -1.1577 \pm i(10.1) \quad \text{(short-period)}$$

Because the real parts of eigenvalues are negative, there are no unstable modes. From the eigenvalues, the period, time to half amplitude, and number of cycles to half amplitude are readily obtained in Table 4.5 - 1.

Table 4.5 - 1 Time to half amplitude, period and number of cycles to half-amplitude

Long-period	Short-period
$t_{1/2} = 0.693/|\eta| = (0.693/0.0066)$ s $= 105$ s;	$t_{1/2} = 0.693/|\eta| = (0.693/1.1577)$ s $= 0.599$ s;
period: $2\pi/\omega = (2\pi/0.0215)$ s $= 292$ s;	period: $2\pi/\omega = (2\pi/10.1)$ s $= 0.622$ s;
number of cycles to half-amplitude:	number of cycles to half-amplitude:
$N_{1/2} = 0.11 \dfrac{\omega}{|\eta|} = 0.11 \times \dfrac{0.0215}{0.0066} = 0.358$ (cycles)	$N_{1/2} = 0.11 \dfrac{\omega}{|\eta|} = 0.11 \times \dfrac{10.1}{1.1577} = 0.96$ (cycles)

Example 4.5 - 2: An anti-tank missile is cruising in horizontal flight at flight velocity $V = 118$ m/s. The homogeneous equation is given by

CHAPTER 4 Longitudinal Motion

$$\begin{bmatrix} \Delta \dot{u} \\ \Delta \dot{q} \\ \Delta \dot{\alpha} \\ \Delta \dot{\theta} \end{bmatrix} = \begin{bmatrix} -0.1102 & 0 & -17.256+9.786 & -9.786 \\ -0.000487 & -1.3415 & -126.78 & 0 \\ -0.00162 & 1 & -1.4764+0.00582 & -0.00582 \\ 0 & 1 & 0 & 0 \end{bmatrix} \begin{bmatrix} \Delta u \\ \Delta q \\ \Delta \alpha \\ \Delta \theta \end{bmatrix}$$

MATLAB program 4.5-2 generates the coefficients of the characteristic polynomial and eigenvalues of the coefficient matrix.

MATLAB program 4.5-2

```
>>format long
>>e=[-0.1102 0 -17.256+9.786 -9.786;
    -0.000487 -1.3415 -126.78 0;
    -0.00162 1 -1.4764+0.00582 -0.00582;
    0 1 0 0];
>> poly(e)
ans =
    1.0000    2.9223   129.0506   13.4261    1.9216
>> eig(e)
ans =
   -1.4092+11.2586i
   -1.4092-11.2586i
   -0.0520+0.1106i
   -0.0520-0.1106i
```

Thus the characteristic equation is directly written as

$$\lambda^4 + 2.9223\lambda^3 + 129.0506\lambda^2 + 13.4261\lambda + 1.9216 = 0$$

The roots of the characteristic equation are

$$\lambda_{1,2} = -0.0520 \pm i(0.1106) \quad \text{(long-period)}$$
$$\lambda_{3,4} = -1.4092 \pm i(11.2586) \quad \text{(short-period)}$$

Since the real parts of eigenvalues are negative, there are no unstable modes. From the eigenvalues, the period, time to half amplitude, and number of cycles to half amplitude are readily obtained in Table 4.5-2.

4.5 Solution of the Longitudinal Equations (Control Surface Locked)

Table 4.5-2 Time to half amplitude, period and number of cycles to half-amplitude

Long-period	Short-period
$t_{1/2}=0.693/|\eta|=(0.693/0.052)\text{s}=13.32\text{ s}$;	$t_{1/2}=0.693/|\eta|=(0.693/1.4092)\text{s}=0.492\text{ s}$;
period: $2\pi/\omega=(2\pi/0.1106)\text{s}=56.81\text{ s}$;	period: $2\pi/\omega=(2\pi/11.2586)\text{s}=0.558\text{ s}$;
number of cycles to half-amplitude:	number of cycles to half-amplitude:
$N_{1/2}=0.11\dfrac{\omega}{|\eta|}=0.11\times\dfrac{0.1106}{0.052}=0.234$ (cycles)	$N_{1/2}=0.11\dfrac{\omega}{|\eta|}=0.11\times\dfrac{11.2586}{1.4092}=0.8791$ (cycles)

Example 4.5-3: A shore to ship missile is flying at altitude $H=100$ m with a flight velocity $V=312.7$ m/s. The homogeneous equations is given by

$$\begin{bmatrix}\Delta\dot{u}\\ \Delta\dot{q}\\ \Delta\dot{\alpha}\\ \Delta\dot{\theta}\end{bmatrix}=\begin{bmatrix}-0.00601 & 0 & -0.0647+9.8 & -9.8\\ -0.007697+0.5716\times0.00007107 & -2.125-0.5716 & 0.5716\times0.943-45.397 & 0\\ -0.00007107 & 1 & -0.943 & 0\\ 0 & 1 & 0 & 0\end{bmatrix}\begin{bmatrix}\Delta u\\ \Delta q\\ \Delta\alpha\\ \Delta\theta\end{bmatrix}$$

MATLAB program 4.5-3 generates the coefficients of the characteristic polynomial and eigenvalues of the coefficient matrix.

MATLAB program 4.5-3

`>> a=[-0.00601 0 -0.0647+9.8 -9.8;`
`-0.007697+0.5716*0.00007107 -2.125-0.5716 0.5716*0.943-45.397-0.5716*0;`
`-0.00007107 1 -0.943 0;`
`0 1 0 0];`
`>> poly(a)`
`ans =`
`    1.0000    3.6456   47.4234    0.2862   -0.0395`
`>> eig(a)`
`ans =`
`  -1.8198+6.6401i`
`  -1.8198-6.6401i`
`  -0.0321`
`   0.0260`

For the sake of clarity, the characteristic equation is written as
$$\lambda^4+3.6456\lambda^3+47.423\lambda^2+0.2862\lambda-0.0395=0$$
The roots of the characteristic equation are
$$\lambda_{1,2}=-0.0321,\ 0.026\quad\text{(long-period)}$$
$$\lambda_{3,4}=-1.8198\pm\text{i}(6.6401)\quad\text{(short-period)}$$

CHAPTER 4 Longitudinal Motion

A small positive real root corresponds to an unstable mode. From the eigenvalues, the period, time to half amplitude, and number of cycles to half amplitude can be obtained in Table 4.5-3.

Table 4.5-3 Time to half/double amplitude, period and number of cycles to half amplitude

Long-period	Short-period
$t_{1/2}=0.693/\|\eta\|=(0.693/0.0321)\text{s}=21.59\text{ s};$ $t_2=0.693/\|\eta\|=(0.693/0.026)\text{ s}=26.65\text{ s}$	$t_{1/2}=0.693/\|\eta\|=(0.693/1.8198)\text{ s}=0.38082\text{ s};$ period: $2\pi/\omega=(2\pi/6.6401)\text{ s}=0.946\text{ s};$ number of cycles to half-amplitude: $N_{1/2}=0.11\dfrac{\omega}{\|\eta\|}=0.11\times\dfrac{6.6401}{1.8198}=0.0387\text{(cycles)}$

Example 4.5-4: An unmanned vehicle is flying at altitude $H=18\,000$ m with a flight velocity $V=200$ m/s. The homogeneous equations is given by

$$\begin{bmatrix}\Delta\dot{u}\\ \Delta\dot{q}\\ \Delta\dot{\alpha}\\ \Delta\dot{\theta}\end{bmatrix}=\begin{bmatrix}-0.0074 & 0 & -9.17+9.8 & -9.8\\ -0.001 & -0.28 & -5.9 & 0\\ -0.00066 & 1 & -0.47 & 0\\ 0 & 1 & 0 & 0\end{bmatrix}\begin{bmatrix}\Delta u\\ \Delta q\\ \Delta\alpha\\ \Delta\theta\end{bmatrix}$$

MATLAB program 4.5-4 generates the coefficients of the characteristic polynomial and eigenvalues of the coefficient matrix.

```
MATLAB program 4.5-4

>> a=[-0.0074 0 -9.17+9.8 -9.8;
     -0.001 -0.28 -5.9 0;
     -0.00066 1 -0.47 0;
     0 1 0 0];
>> poly(a)
ans =
    1.0000    0.7574    6.0376    0.0356    0.0336
>> eig(a)
ans =
   -0.3761+2.4262i
   -0.3761-2.4262i
   -0.0026+0.0746i
   -0.0026-0.0746i
```

For the sake of clarity, the characteristic equation is written as
$$\lambda^4 + 0.7574\lambda^3 + 6.0376\lambda^2 + 0.0356\lambda + 0.0336 = 0$$
The roots of the characteristic equation are
$$\lambda_{1,2} = -0.0026 \pm i(0.0746) \quad \text{(long-period)}$$
$$\lambda_{3,4} = -0.376 \pm i(2.426) \quad \text{(short-period)}$$

Since the real parts of eigenvalues are negative, there are no unstable modes. From the eigenvalues, the period, time to half amplitude, and number of cycles to half amplitude can be obtained in Table 4.5-4.

Table 4.5-4 Time to half amplitude, period and number of cycles to half-amplitude

Long-period	Short-period
$t_{1/2} = 0.693/|\eta| = (0.693/0.0026)$ s $= 266.5$ s; period: $2\pi/\omega = (2\pi/0.0746)$ s $= 84.23$ s; number of cycles to half-amplitude: $N_{1/2} = 0.11 \frac{\omega}{|\eta|} = 0.11 \times \frac{0.0746}{0.0026} = 3.16$ (cycles)	$t_{1/2} = 0.693/|\eta| = (0.693/0.376)$ s $= 1.843$ s; period: $2\pi/\omega = (2\pi/2.426)$ s $= 2.59$ s; number of cycles to half-amplitude: $N_{1/2} = 0.11 \frac{\omega}{|\eta|} = 0.11 \times \frac{2.426}{0.376} = 0.711$ (cycles)

Examinations of the four examples presents that although missiles or vehicles belong to different types, their eigenvalues have the same characteristics, i.e. magnitude of a pair of eigenvalues is further greater than that of the other pair.

4.6 Transient Response of Vehicle

Example 4.6-1: An anti-tank missile state equation is given by

$$\begin{bmatrix} \Delta\dot{u} \\ \Delta\dot{q} \\ \Delta\dot{\alpha} \\ \Delta\dot{\theta} \end{bmatrix} = \begin{bmatrix} -0.1102 & 0 & -17.256+9.786 & -9.786 \\ -0.000487 & -1.3415 & -126.78 & 0 \\ -0.00162 & 1 & -1.4764+0.00582 & -0.00582 \\ 0 & 1 & 0 & 0 \end{bmatrix} \begin{bmatrix} \Delta u \\ \Delta q \\ \Delta\alpha \\ \Delta\theta \end{bmatrix} + \begin{bmatrix} 0 \\ -16.508 \\ -0.01935 \\ 0 \end{bmatrix} \Delta\delta_e$$

MATLAB program 4.6-1 produces the zero-input response of the system with the initial condition $\Delta\alpha_0 = 2°$. The resulting plot is shown in Figure 4.6-1.

MATLAB program 4.6 – 1

```
>> A=[-0.1102 0 -17.256+9.786 -9.786;
      -0.000487 -1.3415 -126.78 0;
      -0.00162 1 -1.4764+0.00582 -0.00582;
      0 1 0 0];
>> B=[0;-16.508;-0.01935;0];
>> C=[1 0 0 0;0 1 0 0;0 0 1 0;0 0 0 1];
>> D=[0;0;0;0];
>> x0=[0;0;2/57.3;0];
>> initial(A,B,C,D,x0,40);
```

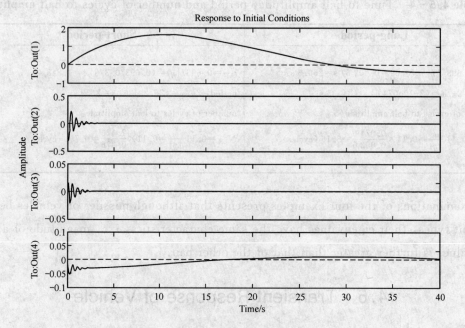

Figure 4.6 – 1 Zero-input response for 40 seconds

The zero-input response for 2 seconds will be obtained with the following MATLAB command

```
>> initial(A,B,C,D,x0,2);
```

The plot obtained by using this command is shown in Figure 4.6 – 2.

In a similar manner, to obtain the step response, enter the following MATLAB commands

```
>> T=0:0.05:3;
>> step(A,B,C,D,1,T);
```

into the computer. The resulting plot is shown in Figure 4.6 – 3.

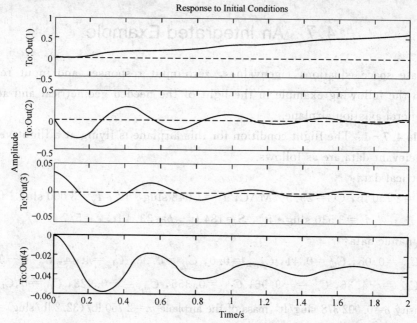

Figure 4.6 – 2 Zero-input response for 2 seconds

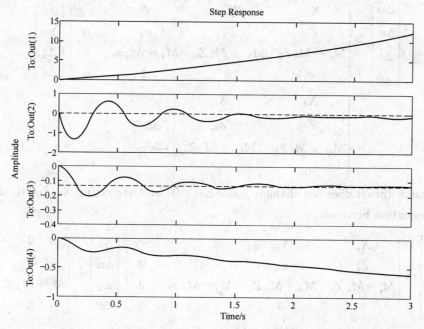

Figure 4.6 – 3 Step response

CHAPTER 4 Longitudinal Motion

In the figures, the out(1), out(2), out(3), and out(4) are Δu, Δq, $\Delta \alpha$, $\Delta \theta$ respectively.

4.7 An Integrated Example

The state space equation, eigenvalues, zero-input response, and input response are presented in the following example in the light of the needed geometrical and aerodynamic data of a general aviation airplane.

Example 4.7-1: The flight condition for this airplane is flying at a flight velocity $u_0 = 176$ ft/s. Relevant data are as follows.

Geometrical data:

$$W = 2\,750 \text{ lb}, \text{ CG} = 29.5\%\text{MAC}, I_x = 1\,048 \text{ slug} \cdot \text{ft}^2, I_y = 3\,000 \text{ slug} \cdot \text{ft}^2$$

$$I_z = 3\,530 \text{ slug} \cdot \text{ft}^2, S = 184 \text{ ft}^2, b = 33.4 \text{ ft}, \bar{c} = 5.7 \text{ ft}$$

Aerodynamic data:

$$C_{D_0} = 0.05, C_{L_0} = 0.41, C_{m,Ma} = 0.0, C_{D_\alpha} = 0.33, C_{L_\alpha} = 4.44, C_{m_\alpha} = -0.683$$

$$C_{m_{\dot\alpha}} = -4.36, C_{m_q} = -9.96, C_{L_{\delta_e}} = 0.355, C_{m_{\delta_e}} = -0.923, C_{D_u} = 0, C_{L_u} = 0$$

Air density $\rho = 0.002\,378$ slug/ft^3, mass of the airplane $m = 2\,750 \text{ lb}/(32.2 \text{ lb/slug}) = 85.4$ slug.

The state equation (4.2-1) of a vehicle is rewritten as

$$\begin{bmatrix} \Delta \dot u \\ \Delta \dot w \\ \Delta \dot q \\ \Delta \dot \theta \end{bmatrix} = \begin{bmatrix} X_u & X_w & 0 & -g \\ Z_u & Z_w & u_0 & 0 \\ M_u + M_{\dot w} Z_u & M_w + M_{\dot w} Z_w & M_q + M_{\dot w} u_0 & 0 \\ 0 & 0 & 1 & 0 \end{bmatrix} \begin{bmatrix} \Delta u \\ \Delta w \\ \Delta q \\ \Delta \theta \end{bmatrix} +$$

$$\begin{bmatrix} X_{\delta_e} & X_{\delta_T} \\ Z_{\delta_e} & Z_{\delta_T} \\ M_{\delta_e} + M_{\dot w} Z_{\delta_e} & M_{\delta_T} + M_{\dot w} Z_{\delta_T} \\ 0 & 0 \end{bmatrix} \begin{bmatrix} \Delta \delta_e \\ \Delta \delta_T \end{bmatrix}$$

Supposing thrust does not change, i.e., $\Delta \delta_T = 0$, and neglecting X_{δ_e} due to small value, the above equation becomes

$$\begin{bmatrix} \Delta \dot u \\ \Delta \dot w \\ \Delta \dot q \\ \Delta \dot \theta \end{bmatrix} = \begin{bmatrix} X_u & X_w & 0 & -g \\ Z_u & Z_w & u_0 & 0 \\ M_u + M_{\dot w} Z_u & M_w + M_{\dot w} Z_w & M_q + M_{\dot w} u_0 & 0 \\ 0 & 0 & 1 & 0 \end{bmatrix} \begin{bmatrix} \Delta u \\ \Delta w \\ \Delta q \\ \Delta \theta \end{bmatrix} + \begin{bmatrix} 0 \\ Z_{\delta_e} \\ M_\delta + M_{\dot w} Z_{\delta_e} \\ 0 \end{bmatrix} \Delta \delta_e$$

(4.7-1)

In order to estimate longitudinal stability derivatives, the dynamic pressure Q, and the terms QS, $QS\bar{c}$, and $\bar{c}/2u_0$ are computed firstly as follows:

$$Q = \frac{1}{2}\rho u_0^2 = (0.5)(0.002\ 378\ \text{slug/ft}^3)(176\ \text{ft/s})^2 = 36.83\ \text{lb/ft}^2$$

$$QS = (36.83\ \text{lb/ft}^2)(184\ \text{ft}^2) = 6\ 776.72\ \text{lb}$$

$$QS\bar{c} = (6\ 776.72\ \text{lb})(5.7\ \text{ft}) = 38\ 627.30\ \text{ft}\cdot\text{lb}$$

$$(\bar{c}/2u_0) = (5.7\ \text{ft})/(2\times 176\ \text{ft/s}) = 0.016\ 2\ \text{s}$$

The longitudinal derivatives can be evaluated from the formulas in Table 4.1-1.
The stability derivatives with respect to u

$$X_u = -(C_{D_u} + 2C_{D_0})QS/(u_0 m) =$$
$$-(0.0 + 2\times 0.05)(6\ 776.72\ \text{lb})/[(176\ \text{ft/s})(85.4\ \text{slug})] = -0.045\ 1\ \text{s}^{-1}$$

$$Z_u = -(C_{L_u} + 2C_{L_0})QS/(u_0 m) =$$
$$-(0.0 + 2\times 0.41)(6\ 776.72\ \text{lb})/[(176\ \text{ft/s})(85.4\ \text{slug})] = -0.370\ \text{s}^{-1}$$

$$M_u = 0$$

The stability derivatives with respect to w

$$X_w = -(C_{D_\alpha} - C_{L_0})QS/(u_0 m) =$$
$$-(0.33 - 0.41)(6\ 776.72\ \text{lb})/[(176\ \text{ft/s})(85.4\ \text{slug})] = 0.036\ 1\ \text{s}^{-1}$$

$$Z_w = -(C_{L_\alpha} + C_{D_0})QS/(u_0 m) =$$
$$-(4.44 + 0.05)(6\ 776.72\ \text{lb})/[(176\ \text{ft/s})(85.4\ \text{slug})] = -2.024\ \text{s}^{-1}$$

$$M_w = C_{m_\alpha} QS\bar{c}/(u_0 I_y) =$$
$$(-0.683)(38\ 627.3\ \text{ft}\cdot\text{lb})/[(176\ \text{ft/s})(3\ 000\ \text{slug}\cdot\text{ft}^2)] = -0.05\ (\text{ft}\cdot\text{s})^{-1}$$

The stability derivatives with respect to $\dot{w}$

$$X_{\dot{w}} = 0$$

$$Z_{\dot{w}} = 0$$

$$M_{\dot{w}} = C_{m_{\dot{\alpha}}} \frac{\bar{c}}{2u_0} QS\bar{c}/(u_0 I_y) =$$
$$(-4.36)(0.016\ 2\ \text{s})(38\ 627.3\ \text{ft}\cdot\text{lb})/[(176\ \text{ft/s})(3\ 000\ \text{slug}\cdot\text{ft}^2)] =$$
$$-0.005\ 2\ \text{ft}^{-1}$$

The stability derivatives with respect to q

$$X_q = 0$$

$$Z_q = 0$$

$$M_q = C_{m_q} \frac{\bar{c}}{2u_0} QS\bar{c}/I_y =$$
$$(-9.96)(0.016\ 2\ \text{s})(38\ 627.3\ \text{ft}\cdot\text{lb})/(3\ 000\ \text{slug}\cdot\text{ft}^2) = -2.078\ \text{s}^{-1}$$

The stability derivatives with respect to δ_e

CHAPTER 4 Longitudinal Motion

$$Z_{\delta_e} = \frac{C_{Z_{\delta_e}} QS}{m} = -\frac{C_{L_{\delta_e}} QS}{m} = -\frac{0.355 \times 6\,776.72}{85.4} \text{ ft/s}^2 = -28.17 \text{ ft/s}^2$$

$$M_\delta + M_{\dot{w}} Z_{\delta_e} = \frac{C_{m_{\delta_e}}(QS\bar{c})}{I_y} + C_{m_{\dot{\alpha}}} \frac{\bar{c}}{2u_0} \frac{QS\bar{c}}{u_0 I_y} Z_{\delta_e} = \left(C_{m_{\delta_e}} + C_{m_{\dot{\alpha}}} \frac{\bar{c}}{2u_0 u_0} Z_{\delta_e}\right) \frac{QS\bar{c}}{I_y} =$$

$$\left(-0.923 + 4.36 \frac{5.7}{2 \times 176 \times 176} \times 28.17\right) \times \frac{38\,627.3}{3\,000} = -11.74 \text{ s}^{-2}$$

Substituting the above values into the equation (4.7-1) yields state equation as follows:

$$\begin{bmatrix} \Delta \dot{u} \\ \Delta \dot{w} \\ \Delta \dot{q} \\ \Delta \dot{\theta} \end{bmatrix} = \begin{bmatrix} -0.045 & 0.036\,1 & 0.000 & -32.2 \\ -0.37 & -2.024 & 176 & 0.000 \\ 0.001\,924 & -0.039\,5 & -2.999\,932 & 0.000 \\ 0.000 & 0.000 & 1.000 & 0.000 \end{bmatrix} \begin{bmatrix} \Delta u \\ \Delta w \\ \Delta q \\ \Delta \theta \end{bmatrix} + \begin{bmatrix} 0 \\ -28.17 \\ -11.74 \\ 0 \end{bmatrix} \delta_e$$

(4.7-2)

In view of $\Delta w = u_0 \Delta \alpha$, then

$$\begin{bmatrix} \Delta \dot{u} \\ \Delta \dot{\alpha} \\ \Delta \dot{q} \\ \Delta \dot{\theta} \end{bmatrix} = \begin{bmatrix} -0.045 & 0.036\,1 \times 176 & 0 & -32.2 \\ \dfrac{-0.37}{176} & -2.024 & 1 & 0 \\ 0.001\,924 & -0.039\,5 \times 176 & -2.999\,932 & 0 \\ 0 & 0 & 1 & 0 \end{bmatrix} \begin{bmatrix} \Delta u \\ \Delta \alpha \\ \Delta q \\ \Delta \theta \end{bmatrix} + \begin{bmatrix} 0 \\ -\dfrac{28.17}{176} \\ -11.74 \\ 0 \end{bmatrix} \Delta \delta_e$$

MATLAB program 4.7-1 generates the coefficients of the characteristic polynomial and eigenvalues of the system.

MATLAB program 4.7-1

```
>> A=[-0.045 0.0361*176 0 -32.2;
     -0.37/176 -2.024 1 0;
     0.001924 -0.0395*176 -2.999932 0;
     0 0 1 0];
>> poly(A)
ans =
   1.0000   5.0689   13.2633   0.6759   0.5960
>> eig(A)
ans =
   -2.5174+2.5900i
   -2.5174-2.5900i
   -0.0171+0.2131i
   -0.0171-0.2131i
```

For the sake of clarity, the characteristic equation is written as
$$\lambda^4 + 5.0689\lambda^3 + 13.2633\lambda^2 + 0.6759\lambda + 0.5960 = 0$$
The roots of the characteristic equation are
$$\lambda_{1,2} = -0.0171 \pm i(0.213) \quad \text{(long-period)}$$
$$\lambda_{3,4} = -2.5174 \pm i(2.59) \quad \text{(short-period)}$$
From the eigenvalues, the period, time to half amplitude, and number of cycles to half amplitude can be obtained in Table 4.7-1.

Table 4.7-1 Time to half amplitude, period and number of cycles to half-amplitude

Long-period	Short-period
$t_{1/2} = 0.69/|\eta| = 0.69/0.0171 \text{ s} = 40.35 \text{ s}$;	$t_{1/2} = 0.69/|\eta| = 0.69/2.5174 \text{ s} = 0.274 \text{ s}$;
period: $2\pi/\omega = 2\pi/0.213 \text{ s} = 29.5 \text{ s}$;	period: $2\pi/\omega = 2\pi/2.59 \text{ s} = 2.43 \text{ s}$;
number of cycles to half-amplitude:	number of cycles to half-amplitude:
$N_{1/2} = 0.11 \frac{\omega}{|\eta|} = 0.11 \times \frac{0.213}{0.0171} = 1.37 \text{ (cycles)}$	$N_{1/2} = 0.11 \frac{\omega}{|\eta|} = 0.11 \times \frac{2.59}{2.5174} = 0.113 \text{ (cycles)}$

The eignvalues are also estimated by means of the foregoing approximations.

The long-period motion approximation equation (4.3-18) is rewritten as
$$\lambda^2 - X_u\lambda - \frac{Z_u g}{u_0} = 0$$
Substituting the stability derivatives values, X_u, and Z_u yields
$$\lambda^2 + 0.045\lambda - \frac{-0.37 \times 32.21}{176} = 0$$
MATLAB program 4.7-2 generates the roots of the equation.

MATLAB program 4.7-2

```
>> p=[1 0.045 0.37*32.21/176];
>> roots(p)
ans =
    -0.0225+0.2592i
    -0.0225-0.2592i
```

That is
$$\lambda_{1,2} = -0.0225 \pm i(0.2592) \quad \text{(long-period)}$$
Compared with the accurate values, the long-period approximate eignvalues have some errors.

CHAPTER 4   Longitudinal Motion

The short-period motion equation (4.3-4) is rewritten as

$$\lambda^2 - \left(M_q + M_{\dot{\alpha}} + \frac{Z_\alpha}{u_0}\right)\lambda + M_q\frac{Z_\alpha}{u_0} - M_\alpha = 0$$

Substituting relevant values yields

$$\lambda^2 - (-2.2078 - 0.0052 \times 176 - 2.024)\lambda - 2.078 \times$$
$$(-2.024) - (-0.05) \times 176 = 0$$

MATLAB program 4.7-3 generates the roots of the equation.

MATLAB program 4.7-3

```
>> p=[1 2.078+176*0.0052+2.024 2.078*2.024+0.05*176];
>> roots(p)
ans =
    -2.5086+2.5909i
    -2.5086-2.5909i
```

That is

$$\lambda_{3,4} = -2.5086 \pm i(2.5909) \quad \text{(short-period)}$$

Compared with the accurate values, the short-period values are very satisfactory.

The transfer function will be obtained through MATLAB program 4.7-4.

MATLAB program 4.7-4

```
>> A=[-0.045 0.0361*176 0 -32.2;
     -0.37/176 -2.024 1 0;
     0.001924 -0.0395*176 -2.999932 0;
     0 0 1 0];
>> B=[0;-28.17/176;-11.74;0];
>> C=[1 0 0 0;0 1 0 0;0 0 1 0;0 0 0 1];
>> D=[0;0;0;0];
>> [num,den]=ss2tf(A,B,C,D,1)
num =
    0   -0.0000    -1.0169   300.3860   729.2992
    0   -0.1601   -12.2274    -0.5499    -0.8046
    0  -11.7400   -23.1773    -1.1780    -0.0000
    0   -0.0000   -11.7400   -23.1773    -1.1780
den =
    1.0000    5.0689   13.2633    0.6759    0.5960
```

4.7 An Integrated Example

As an example, the transfer function $\Delta\theta(s)/\Delta\delta_e(s)$ is

$$\frac{\Delta\theta(s)}{\Delta\delta_e(s)} = -\frac{11.74s^2 + 23.1773s + 1.1780}{s^4 + 5.0689s^3 + 13.2633s^2 + 0.6759s + 0.596}$$

To obtain the zero-input response with the initial condition $\Delta\alpha_0 = 2°$, enter the following MATLAB commands

```
>> x0=[0;2/57.3;0;0];
>> initial(A,B,C,D,x0,20);
```

into the computer. The resulting plot is shown in Figure 4.7-1.

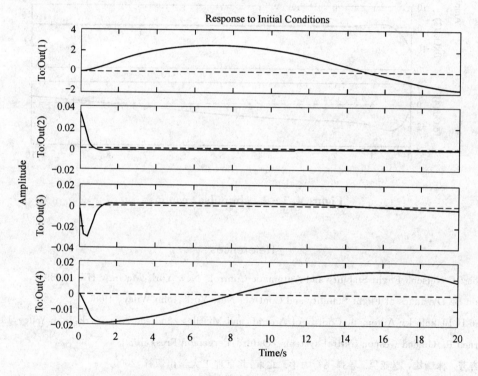

Figure 4.7-1   Zero-input response

In a similar manner, obtain the impulse-input response by the following MATLAB commands.

```
>> T=0:0.05:10;
>> impulse(A,B,C,D,1,T);
```

Figure 4.7-2 shows the impulse-input response.

CHAPTER 4 Longitudinal Motion

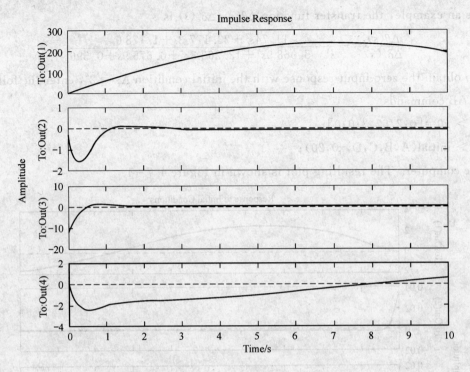

Figure 4.7 – 2 Impulse response

References

[1] Robert C Nelson. Flight Stability and Automatic Control. New York:McGraw-Hill, 1989.

[2] Etkin B. Dynamics of Flight Stability and Control. New York:John Wiley, 1996.

[3] John H Blakelock. Automatic Control of Aircraft and Missiles. 2nd ed. New York:John Wiley,1991.

[4] Garnell P. Guided weapon control systems. Oxford:Pergamon Press,1980.

[5] 钱杏芳,林瑞雄,赵亚男. 导弹飞行力学. 北京:北京理工大学出版社,2000.

CHAPTER 5
Lateral Motion

This chapter mainly discusses mathematical models about lateral dynamic motion to build a foundation for lateral motion control system design and analysis. Based on the results of Chapter 3, the lateral disturbance motion state equations and transfer functions are derived. The lateral disturbance motion includes three modes: ① roll mode, ② Dutch roll mode, and ③ spiral mode. Making some assumptions, approximate relationships of these modes are derived. In the chapter, some examples about lateral disturbance motion are illustrated. For missiles with two symmetrical planes, the reduction of small disturbance motion equations is presented in this chapter.

5.1 State Variable Representation of the Linearized Lateral Equations

For convenience, the lateral small disturbance motion equations (3.4 - 21) is rewritten as

$$\left.\begin{aligned}\left(\frac{\mathrm{d}}{\mathrm{d}t}-Y_v\right)\Delta v - Y_p\Delta p + (u_0-Y_r)\Delta r - (g\cos\theta_0)\Delta\phi &= Y_{\delta_r}\Delta\delta_r \\ -L_v\Delta v + \left(\frac{\mathrm{d}}{\mathrm{d}t}-L_p\right)\Delta p - \left(\frac{I_{xx}}{I_x}\frac{\mathrm{d}}{\mathrm{d}t}+L_r\right)\Delta r &= L_{\delta_a}\Delta\delta_a + Z_{\delta_r}\Delta\delta_r \\ -N_v\Delta v - \left(\frac{I_{xx}}{I_z}\frac{\mathrm{d}}{\mathrm{d}t}+N_p\right)\Delta p + \left(\frac{\mathrm{d}}{\mathrm{d}t}-N_r\right)\Delta r &= N_{\delta_a}\Delta\delta_a + N_{\delta_r}\Delta\delta_r \end{aligned}\right\}$$

In order to obtain state space equations, it is necessary to make manipulations as follows:

① For the first equation of the above equations, the $\frac{\mathrm{d}\Delta v}{\mathrm{d}t}$ is retained, while the other leftover terms are moved to the right side of the equation,

CHAPTER 5 Lateral Motion

② The $\dfrac{d\Delta r}{dt}$ is eliminated through the third equation multiplied by $\dfrac{I_{xz}}{I_x}$ plus the second equation to obtain $\dfrac{d\Delta p}{dt}$ equation,

③ The $\dfrac{d\Delta p}{dt}$ is eliminated through the second equation multiplied by $\dfrac{I_{xz}}{I_z}$ plus the third equation to obtain $\dfrac{d\Delta r}{dt}$ equation.

After the above manipulations, and adding an equation (3.4-22) (i.e. $\Delta\dot\phi = \Delta p + \Delta r \tan\theta_0$), the equations become

$$\begin{bmatrix}\Delta\dot v\\ \Delta\dot p\\ \Delta\dot r\\ \Delta\dot\phi\end{bmatrix} = \begin{bmatrix} Y_v & Y_p & -(u_0-Y_r) & g\cos\theta_0 \\ L_v^* + \dfrac{I_{xz}}{I_x}N_v^* & L_p^* + \dfrac{I_{xz}}{I_x}N_p^* & L_r^* + \dfrac{I_{xz}}{I_x}N_r^* & 0 \\ N_v^* + \dfrac{I_{xz}}{I_z}L_v^* & N_p^* + \dfrac{I_{xz}}{I_z}L_p^* & N_r^* + \dfrac{I_{xz}}{I_z}L_v^* & 0 \\ 0 & 1 & \tan\theta_0 & 0 \end{bmatrix} \begin{bmatrix}\Delta v\\ \Delta p\\ \Delta r\\ \Delta\phi\end{bmatrix} +$$

$$\begin{bmatrix} 0 & Y_{\delta_r} \\ L_{\delta_a}^* + \dfrac{I_{xz}}{I_x}N_{\delta_a}^* & L_{\delta_r}^* + \dfrac{I_{xz}}{I_x}N_{\delta_r}^* \\ N_{\delta_a}^* + \dfrac{I_{xz}}{I_z}L_{\delta_a}^* & N_{\delta_r}^* + \dfrac{I_{xz}}{I_z}L_{\delta_r}^* \\ 0 & 0 \end{bmatrix} \begin{bmatrix}\Delta\delta_a\\ \Delta\delta_r\end{bmatrix} \qquad (5.1-1)$$

The symbols with asterisk are defined as

$$L_v^* = \dfrac{L_v}{1-[I_{xz}^2/(I_xI_z)]}, \quad N_v^* = \dfrac{N_v}{1-[I_{xz}^2/(I_xI_z)]}, \quad \text{etc.}$$

If the product of inertia $I_{xz}=0$, and $\theta_0=0$, the state equation is simplified to the following form

$$\begin{bmatrix}\Delta\dot v\\ \Delta\dot p\\ \Delta\dot r\\ \Delta\dot\phi\end{bmatrix} = \begin{bmatrix} Y_v & Y_p & -(u_0-Y_r) & g \\ L_v & L_p & L_r & 0 \\ N_v & N_p & N_r & 0 \\ 0 & 1 & 0 & 0 \end{bmatrix} \begin{bmatrix}\Delta v\\ \Delta p\\ \Delta r\\ \Delta\phi\end{bmatrix} + \begin{bmatrix} 0 & Y_{\delta_r} \\ L_{\delta_a} & L_{\delta_r} \\ N_{\delta_a} & N_{\delta_r} \\ 0 & 0 \end{bmatrix} \begin{bmatrix}\Delta\delta_a\\ \Delta\delta_r\end{bmatrix} \qquad (5.1-2)$$

Sometimes, use the sideslip angle $\Delta\beta$ instead of the side velocity Δv. In view of $\Delta\beta \approx \Delta v/u_0$, the equation (5.1-2) becomes

$$\begin{bmatrix} \Delta\dot\beta \\ \Delta\dot p \\ \Delta\dot r \\ \Delta\dot\phi \end{bmatrix} = \begin{bmatrix} \dfrac{Y_\beta}{u_0} & \dfrac{Y_p}{u_0} & -\left(1-\dfrac{Y_r}{u_0}\right) & \dfrac{g}{u_0} \\ L_\beta & L_p & L_r & 0 \\ N_\beta & N_p & N_r & 0 \\ 0 & 1 & 0 & 0 \end{bmatrix} \begin{bmatrix} \Delta\beta \\ \Delta p \\ \Delta r \\ \Delta\phi \end{bmatrix} + \begin{bmatrix} 0 & \dfrac{Y_{\delta_r}}{u_0} \\ L_{\delta_a} & L_{\delta_r} \\ N_{\delta_a} & N_{\delta_r} \\ 0 & 0 \end{bmatrix} \begin{bmatrix} \Delta\delta_a \\ \Delta\delta_r \end{bmatrix} \quad (5.1-3)$$

Table 5.1-1 shows calculating expressions of the lateral dimensional stability derivatives.

Table 5.1-1 Lateral directional derivatives

Lateral directional derivatives	Lateral directional derivatives
$Y_\beta = \dfrac{QSC_{y_\beta}}{m}$ (ft/s² or m/s²)	$N_\beta = \dfrac{QSbC_{n_\beta}}{I_z}$ (s^{-2})
$L_\beta = \dfrac{QSbC_{l_\beta}}{I_x}$ (s^{-2})	$Y_p = \dfrac{QSC_{y_p}}{2mu_0}$ (ft/s or m/s)
$N_p = \dfrac{QSb^2 C_{n_p}}{2I_z u_0}$ (s^{-1})	$L_p = \dfrac{QSb^2 C_{l_p}}{2I_x u_0}$ (s^{-1})
$Y_r = \dfrac{QSC_{y_r}}{2mu_0}$ (ft/s or m/s)	$N_r = \dfrac{QSb^2 C_{n_r}}{2I_z u_0}$ (s^{-1})
$L_r = \dfrac{QSb^2 C_{l_r}}{2I_x u_0}$ (s^{-1})	—
$Y_{\delta_a} = \dfrac{QSC_{y_{\delta_a}}}{m}$ (ft/s² or m/s²)	$N_{\delta_a} = \dfrac{QSbC_{n_{\delta_a}}}{I_z}$ (s^{-2})
$L_{\delta_a} = \dfrac{QSbC_{l_{\delta_a}}}{I_x}$ (s^{-2})	$Y_{\delta_r} = \dfrac{QSC_{y_{\delta_r}}}{m}$ (ft/s² or m/s²)
$N_{\delta_r} = \dfrac{QSbC_{n_{\delta_r}}}{I_z}$ (s^{-2})	$L_{\delta_r} = \dfrac{QSbC_{l_{\delta_r}}}{I_x}$ (s^{-2})

5.2 Lateral Transfer Functions

Taking Laplace transformation of the state equation, and then using Cramer's rule, can obtain single input single output (SISO) transfer functions.

Taking Laplace transformation of the state equation (5.1-3) yields

CHAPTER 5 Lateral Motion

$$\begin{bmatrix} s-\dfrac{Y_\beta}{u_0} & -\dfrac{Y_p}{u_0} & \left(1-\dfrac{Y_r}{u_0}\right) & -\dfrac{g}{u_0} \\ -L_v & s-L_p & -L_r & 0 \\ -N_v & -N_p & s-N_r & 0 \\ 0 & -1 & 0 & s \end{bmatrix} \begin{bmatrix} \Delta\beta(s) \\ \Delta p(s) \\ \Delta r(s) \\ \Delta\phi(s) \end{bmatrix} = \begin{bmatrix} 0 & \dfrac{Y_{\delta_r}}{u_0} \\ L_{\delta_a} & L_{\delta_r} \\ N_{\delta_a} & N_{\delta_r} \\ 0 & 0 \end{bmatrix} \begin{bmatrix} \Delta\delta_a(s) \\ \Delta\delta_r(s) \end{bmatrix} \quad (5.2-1)$$

Let $\Delta\delta_a = 0$, and apply Cramer's rule so as to obtain the transfer functions of $\Delta\delta_r$ input $\Delta\beta(s), \Delta\phi(s)$, and $\Delta r(s)$ outputs as follows:

$$\dfrac{\Delta\beta(s)}{\Delta\delta_r(s)} = \dfrac{N^\beta_{\delta_r}}{\Delta_{\text{lat}}} = \dfrac{\begin{vmatrix} \dfrac{Y_{\delta_r}}{u_0} & -\dfrac{Y_p}{u_0} & \left(1-\dfrac{Y_r}{u_0}\right) & -\dfrac{g}{u_0} \\ L_{\delta_r} & s-L_p & -L_r & 0 \\ N_{\delta_r} & -N_p & s-N_r & 0 \\ 0 & -1 & 0 & s \end{vmatrix}}{\begin{vmatrix} s-\dfrac{Y_\beta}{u_0} & -\dfrac{Y_p}{u_0} & \left(1-\dfrac{Y_r}{u_0}\right) & -\dfrac{g}{u_0} \\ -L_v & s-L_p & -L_r & 0 \\ -N_v & -N_p & s-N_r & 0 \\ 0 & -1 & 0 & s \end{vmatrix}} =$$

$$\dfrac{A_{\beta r}\left(s+\dfrac{1}{T_{\beta r1}}\right)\left(s+\dfrac{1}{T_{\beta r2}}\right)\left(s+\dfrac{1}{T_{\beta r3}}\right)}{\left(s+\dfrac{1}{T_R}\right)\left(s+\dfrac{1}{T_S}\right)(s^2+2\zeta_D\omega_D s+\omega_D^2)} =$$

$$\dfrac{K_{\beta r}(T_{\beta r1}s+1)(T_{\beta r2}s+1)(T_{\beta r3}s+1)}{(T_R s+1)(T_S s+1)(T_D^2 s^2+2\zeta_D T_D s+1)} \quad (5.2-2)$$

where

$N^\beta_{\delta_r} = $ numerator determinant of the transfer function $\Delta\beta/\Delta\delta_r$,

$\Delta_{\text{lat}} = $ lateral motion characteristic determinant,

$A_{\beta r} = $ gain of the transfer function $\Delta\beta/\Delta\delta_r$,

$T_{\beta r1}, T_{\beta r2}, T_{\beta r3} = $ numerator time constants

$T_R = $ rolling mode time constant,

$T_S = $ spiral mode time constant,

$\zeta_D = $ Dutch roll mode damping ratio,

$\omega_D = $ Dutch roll mode natural frequency,

$T_D = 1/\omega_D$, Dutch roll mode time constant,

$K_{\beta r} = $ transfer coefficient of the transfer function $\Delta\beta/\Delta\delta_r$.

$$\frac{\Delta\phi(s)}{\Delta\delta_r(s)} = \frac{N^{\phi}_{\delta_r}}{\Delta_{lat}} = \frac{\begin{vmatrix} s-\dfrac{Y_\beta}{u_0} & -\dfrac{Y_p}{u_0} & \left(1-\dfrac{Y_r}{u_0}\right) & \dfrac{Y_{\delta_r}}{u_0} \\ -L_v & s-L_p & -L_r & L_{\delta_r} \\ -N_v & -N_p & s-N_r & N_{\delta_r} \\ 0 & -1 & 0 & 0 \end{vmatrix}}{\begin{vmatrix} s-\dfrac{Y_\beta}{u_0} & -\dfrac{Y_p}{u_0} & \left(1-\dfrac{Y_r}{u_0}\right) & -\dfrac{g}{u_0} \\ -L_v & s-L_p & -L_r & 0 \\ -N_v & -N_p & s-N_r & 0 \\ 0 & -1 & 0 & s \end{vmatrix}} =$$

$$\frac{A_{\phi r}\left(s+\dfrac{1}{T_{\phi r 1}}\right)\left(s+\dfrac{1}{T_{\phi r 2}}\right)}{\left(s+\dfrac{1}{T_R}\right)\left(s+\dfrac{1}{T_S}\right)(s^2+2\zeta_D\omega_D s+\omega_D^2)} =$$

$$\frac{K_{\phi r}(T_{\phi r 1}s+1)(T_{\phi r 2}s+1)}{(T_R s+1)(T_S s+1)(T_D^2 s^2+2\zeta_D T_D s+1)} \qquad (5.2-3)$$

where

$N^{\phi}_{\delta_r}$ = numerator determinant of the transfer function $\Delta\phi(s)/\Delta\delta_r$,

$A_{\phi r}$ = gain of the transfer function $\Delta\phi/\Delta\delta_r$,

$T_{\phi r 1}$, $T_{\phi r 2}$ = numerator time constants,

$K_{\phi r}$ = transfer coefficient of the transfer function $\Delta\phi/\Delta\delta_r$,

other symbols are defined ibid.

$$\frac{\Delta r(s)}{\Delta\delta_r(s)} = \frac{N^{r}_{\delta_r}}{\Delta_{lat}} = \frac{\begin{vmatrix} s-\dfrac{Y_\beta}{u_0} & -\dfrac{Y_p}{u_0} & \dfrac{Y_{\delta_r}}{u_0} & -\dfrac{g}{u_0} \\ -L_v & s-L_p & L_{\delta_r} & 0 \\ -N_v & -N_p & N_{\delta_r} & 0 \\ 0 & -1 & 0 & 0 \end{vmatrix}}{\begin{vmatrix} s-\dfrac{Y_\beta}{u_0} & -\dfrac{Y_p}{u_0} & \left(1-\dfrac{Y_r}{u_0}\right) & -\dfrac{g}{u_0} \\ -L_v & s-L_p & -L_r & 0 \\ -N_v & -N_p & s-N_r & 0 \\ 0 & -1 & 0 & s \end{vmatrix}} =$$

$$\frac{A_{rr}\left(s+\dfrac{1}{T_{rr1}}\right)(s^2+2\zeta_{rr2}\omega_{rr2}s+\omega_{rr2}^2)}{\left(s+\dfrac{1}{T_R}\right)\left(s+\dfrac{1}{T_S}\right)(s^2+2\zeta_D\omega_D+\omega_D^2)}=$$

$$\frac{K_{rr}(T_{rr1}s+1)(T_{rr2}^2s^2+2\zeta_{rr2}T_{rr2}s+1)}{(T_Rs+1)(T_Ss+1)(T_D^2s^2+2\zeta_DT_D+1)} \tag{5.2-4}$$

where

$N_{\delta_r}^r$ = numerator determinant of the transfer function $\Delta r(s)/\Delta\delta_r$,

A_{rr} = gain of the transfer function $\Delta r/\Delta\delta_r$,

T_{rr1}, T_{rr2} = numerator time constants,

ζ_{rr2} = numerator damping ratio,

ω_{rr2} = numerator natural frequency,

K_{rr} = transfer coefficient the transfer function $\Delta r/\Delta\delta_r$,

other symbols are defined as the foregoing.

In a similar manner, other transfer functions for the $\Delta\delta_a$ input may be obtained. They are

$$\frac{\Delta\beta(s)}{\Delta\delta_a(s)}=\frac{N_{\delta_a}^\beta}{\Delta_{lat}}=\frac{A_{\beta_a}(s^2+2\zeta_{\beta_a}\omega_{\beta_a}s+\omega_{\beta_a}^2)}{\left(s+\dfrac{1}{T_R}\right)\left(s+\dfrac{1}{T_S}\right)(s^2+2\zeta_D\omega_D+\omega_D^2)} \tag{5.2-5}$$

$$\frac{\Delta\phi(s)}{\Delta\delta_a(s)}=\frac{N_{\delta_a}^\phi}{\Delta_{lat}}=\frac{A_{\phi_a}(s^2+2\zeta_{\phi_a}\omega_{\phi_a}s+\omega_{\phi_a}^2)}{\left(s+\dfrac{1}{T_R}\right)\left(s+\dfrac{1}{T_S}\right)(s^2+2\zeta_D\omega_D+\omega_D^2)} \tag{5.2-6}$$

$$\frac{\Delta r(s)}{\Delta\delta_a(s)}=\frac{N_{\delta_a}^r}{\Delta_{lat}}=\frac{A_{r_a}\left(s+\dfrac{1}{T_{ra1}}\right)(s^2+2\zeta_{r_a}\omega_{r_a}s+\omega_{r_a}^2)}{\left(s+\dfrac{1}{T_R}\right)\left(s+\dfrac{1}{T_S}\right)(s^2+2\zeta_D\omega_D+\omega_D^2)} \tag{5.2-7}$$

Here, these symbols definitions are similar to the foregoing.

The denominator of the transfer functions is a characteristic polynomial. Generally, the characteristic polynomial equation has three kinds of roots, which represent rolling mode, spiral mode, and Dutch mode. In a general way, a relatively large magnitude real root corresponds to the rolling mode; a small magnitude real root, which maybe sometimes a positive value, corresponds to the spiral mode; a pair of mid magnitude conjugate complex roots corresponds to the Dutch mode. The typical characteristic roots distribution is presented in Figure 5.2-1.

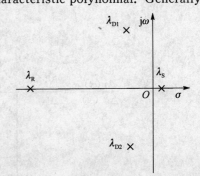

Figure 5.2-1 Typical characteristic roots distribution of lateral motion

The rolling motion presents a rapidly convergent non-period motion. The spiral motion presents a slowly convergent or divergent motion. The Dutch motion shows a lightly damped oscillatory motion, in which the rolling motion is coupling with the yawing motion. The apparent phenomenon of the motion resembles the weaving motion of an ice skater.

5.3 Lateral Approximations

5.3.1 Roll Approximation

1. Roll mode approximation

Consider the second row of the equation (5.1-3). If let $\Delta\beta = \Delta r = 0$, then the homogeneous differential equation is

$$\Delta\dot{p} = L_p \Delta p \qquad (5.3-1)$$

The characteristic equation is

$$\lambda - L_p = 0 \qquad (5.3-2)$$

Therefore,

$$\lambda_R = L_p$$

2. Transfer function of one-degree-of-freedom rolling motion

Supposing that the aileron deflection only produces rolling motion, neglecting sideslip angle and yaw rare, i.e. setting $\Delta\beta = \Delta r = 0$, the second row of the equation (5.1-3) reduces to

$$\Delta\dot{p} = L_p \Delta p + L_{\delta_a} \Delta\delta_a \qquad (5.3-3)$$

where

$$L_p = \frac{\partial L}{\partial p} / I_x,$$

$$L_{\delta_a} = \frac{\partial L}{\partial \delta_a} / I_x.$$

Taking Laplace transformation of the equation (5.3-3) yields

$$(s - L_p)\Delta p(s) = L_{\delta_a} \Delta\delta_a(s)$$

i.e.

CHAPTER 5 Lateral Motion

$$\frac{\Delta p(s)}{\Delta \delta_a(s)} = \frac{L_{\delta_a}}{(s - L_p)} \tag{5.3-4}$$

In view of $\Delta p = \Delta \dot{\varphi}$, then

$$\frac{\Delta \varphi(s)}{\delta_a(s)} = \frac{L_{\delta_a}}{s(s - L_p)} \tag{5.3-5}$$

The reduced transfer function is usually used to choose autopilot parameters for the roll channel of missile.

5.3.2 Effect of Altitude and Airspeed on Roll Mode Characteristic Parameters

From the equation (5.3-4), the roll time constant can be obtained. That is

$$\tau = -\frac{1}{L_p} = -\frac{2I_x u_0}{QSb^2 C_{L_p}} = -\frac{2I_x u_0}{0.5\rho u_0^2 Sb^2 C_{L_p}} = -\frac{4I_x}{Sb^2} \frac{1}{\rho u_0 C_{L_p}} \tag{5.3-6}$$

The expression (5.3-6) presents that the roll time constant is inversely proportional to ρu_0. The time constant decreases as the flight velocity increases. The time constant increases as the altitude increases.

5.3.3 Dutch Roll Approximation

If the Dutch roll mode is considered to primarily consist of sideslipping and yawing motion, then the rolling moment equation can be neglected. With these assumptions, the equation (5.1-3) reduces to

$$\begin{bmatrix} \Delta \dot{\beta} \\ \Delta \dot{r} \end{bmatrix} = \begin{bmatrix} \frac{Y_\beta}{u_0} & -\left(1 - \frac{Y_r}{u_0}\right) \\ N_\beta & N_r \end{bmatrix} \begin{bmatrix} \Delta \beta \\ \Delta r \end{bmatrix} + \begin{bmatrix} 0 & \frac{Y_{\delta_r}}{u_0} \\ N_{\delta_a} & N_{\delta_r} \end{bmatrix} \begin{bmatrix} \Delta \delta_a \\ \Delta \delta_r \end{bmatrix} \tag{5.3-7}$$

Let $\Delta \delta_r = \Delta \delta_a = 0$ to obtain the homogeneous state equations:

$$\begin{bmatrix} \Delta \dot{\beta} \\ \Delta \dot{r} \end{bmatrix} = \begin{bmatrix} \frac{Y_\beta}{u_0} & -\left(1 - \frac{Y_r}{u_0}\right) \\ N_\beta & N_r \end{bmatrix} \begin{bmatrix} \Delta \beta \\ \Delta r \end{bmatrix} \tag{5.3-8}$$

The eigenvalues of the equation (5.3-8) are the solution of the following equation:

$$\begin{vmatrix} \lambda - \frac{Y_\beta}{u_0} & 1 - \frac{Y_r}{u_0} \\ -N_\beta & \lambda - N_r \end{vmatrix} = 0 \tag{5.3-9}$$

Expanding the above determinant yields

$$\lambda^2 - \left(\frac{Y_\beta + u_0 N_r}{u_0}\right)\lambda + \frac{Y_\beta N_r - N_\beta Y_r + u_0 N_\beta}{u_0} = 0 \qquad (5.3-10)$$

The roots of the equation (5.3-10) are given by

$$\lambda_{DR} = \left(N_r + \frac{Y_\beta}{u_0}\right)\Big/2 \pm \left[\left(N_r + \frac{Y_\beta}{u_0}\right)^2 - 4\left(Y_\beta \frac{N_r}{u_0} - Y_r \frac{N_\beta}{u_0} + N_\beta\right)\right]^{\frac{1}{2}}\Big/2 \qquad (5.3-11)$$

The natural frequency and damping ratio are

$$\omega_{DR} = \sqrt{\frac{Y_\beta N_r - N_\beta Y_r + u_0 N_\beta}{u_0}} \qquad (5.3-12)$$

$$\zeta_{DR} = -\frac{1}{2\omega_{DR}}\left(\frac{Y_\beta + u_0 N_r}{u_0}\right) \qquad (5.3-13)$$

The Dutch roll approximation is very useful for obtaining the damping ratio and natural frequency of the Dutch roll mode.

5.3.4 Effect of Altitude and Airspeed on Dutch Roll Mode Characteristic Parameters

Generally, $u_0 N_\beta$ is relatively large; the other terms are small. Thus the natural frequency is expressed as

$$\omega_{DR} = \sqrt{\frac{u_0 N_\beta}{u_0}} = \sqrt{\frac{QSbC_{n_\beta}}{I_z}} = \sqrt{\frac{0.5SbC_{n_\beta}}{I_z}} u_0 \sqrt{\rho} \qquad (5.3-14)$$

The natural frequency is proportional to $u_0 \sqrt{\rho}$. The natural frequency increases as the flight velocity increases. The natural frequency decreases as the altitude increases.

The damping ratio ζ_{DR} is

$$\zeta_{DR} = -\frac{1}{2\sqrt{N_\beta}}\left(\frac{Y_\beta + u_0 N_r}{u_0}\right) = -\frac{0.25SC_{y_\beta} + 0.125Sb^2 C_{n_r}/I_z}{\sqrt{0.5SbC_{n_\beta}/I_z}}\sqrt{\rho} \qquad (5.3-15)$$

The above equation shows that the damping ratio is proportional to $\sqrt{\rho}$. The damping ratio decreases as the altitude increases.

5.3.5 Spiral Approximation

In order to obtain spiral mode, neglecting the side direction force equation and $\Delta\phi$ and letting $\Delta\delta_r = \Delta\delta_a = 0$, the second and third rows of the equation (5.1-3) becomes

$$L_\beta \Delta\beta + L_r \Delta r = 0 \qquad (5.3-16)$$

CHAPTER 5　Lateral Motion

$$\Delta \dot{r} = N_\beta \Delta \beta + N_r \Delta r \tag{5.3-17}$$

Substituting the equation (5.3-16) in the equation (5.3-17) yields

$$\Delta \dot{r} + \frac{L_r N_\beta - L_\beta N_r}{L_\beta} \Delta r = 0 \tag{5.3-18}$$

The characteristic equation of the differential equation is

$$\lambda + \frac{L_r N_\beta - L_\beta N_r}{L_\beta} = 0 \tag{5.3-19}$$

Then,

$$\lambda_{\text{spiral}} = \frac{L_\beta N_r - L_r N_\beta}{L_\beta} \tag{5.3-20}$$

The stability derivatives L_β and N_r are usually negative, whereas N_β and L_r are generally positive. Based on the above sign conditions, the stability condition of the spiral mode is

$$L_\beta N_r - L_r N_\beta > 0 \tag{5.3-21}$$

or

$$L_\beta N_r > L_r N_\beta \tag{5.3-22}$$

Increasing the dihedral effect L_β or the yaw damping N_r can be used to stabilize the spiral mode.

5.4　Effect of Aerodynamic Derivative Variation on Lateral Dynamics Characteristics

From the equation (5.3-6), it can be seen that the time constant of the roll mode is inversely proportional to $|C_{l_p}|$. As the $|C_{l_p}|$ increases, the roll time constant decreases, and the roll motion subsides quickly.

An observation of the equation (5.3-14) shows that the Dutch roll natural frequency is proportional to $\sqrt{C_{n_\beta}}$, that is, increasing C_{n_β} increases the Dutch roll natural frequency. An examination of the equation (5.3-15) presents that the Dutch roll damping ζ_{DR} is proportional to $|C_{y_\beta}|$ and $|C_{n_r}|$. Increasing $|C_{y_\beta}|$ or $|C_{n_r}|$ may increase the natural damping.

From the equation (5.3-21), the spiral mode stability condition can be obtained:

$$C_{l_\beta} C_{n_r} - C_{l_r} C_{n_\beta} > 0 \tag{5.4-1}$$

For spiral stability, $|C_{l_\beta}|$ should be increased.

The most effect of the stability derivatives on the lateral modes is summarized in Table 5.4-1.

Table 5.4 − 1 The most influence of dimensionless stability derivatives on the lateral modes

Aerodynamic derivatives	Quantity most affected	How affected
$\|C_{n_r}\|$	Damping of the Dutch roll, ζ_{DR}	Increase $\|C_{n_r}\|$ to increase the damping
C_{n_β}	Natural frequency of the Dutch roll, ω_{DR}	Increase C_{n_β} to increase the natural frequency
$\|C_{l_p}\|$	Roll subsidence	Increase $\|C_{l_p}\|$ to increase $1/\tau$
$\|C_{l_\beta}\|$	Spiral divergence	Increase $\|C_{l_\beta}\|$ for spiral stability

5.5 Examples of Lateral Motion

Example 5.5 − 1: This example is the same vehicle as the Example 4.7 − 1. Geometrical parameters and flight data are omitted due to the same flight conditions. Relevant lateral data are as follows:

$C_{y_\beta} = -0.564, C_{l_\beta} = -0.074, C_{n_\beta} = 0.071, C_{l_p} = -0.41, C_{n_p} = -0.0575, C_{l_r} = 0.107$

$C_{n_r} = -0.125, C_{l_{\delta_a}} = -0.134, C_{n_{\delta_a}} = -0.0035, C_{y_{\delta_r}} = 0.157, C_{l_{\delta_r}} = 0.107, C_{n_{\delta_r}} = -0.072$

$C_{y_p} = 0, C_{y_r} = 0, C_{y_{\delta_r}} = 0.157, C_{l_{\delta_a}} = -0.134, C_{l_{\delta_r}} = 0.107, C_{n_{\delta_a}} = -0.0035, C_{n_{\delta_r}} = -0.072$

The parameters are evaluated as follows:

$$Q = \frac{1}{2}\rho u_0^2 = (0.5)(0.002\,378\text{ slug/ft}^3)(176\text{ ft/s})^2 = 36.83\text{ lb/ft}^2$$

$$QS = (36.83\text{ lb/ft}^2)(184\text{ ft}^2) = 6\,776.72\text{ lb}$$

$$Y_\beta = \frac{QSC_{y_\beta}}{m} = \frac{6\,776.72 \times (-0.564)}{85.4}\text{ ft/s}^2 = -44.7549\text{ ft/s}^2$$

$$Y_\beta/u_0 = (-44.7549/176)\text{ s}^{-1} = -0.2543\text{ s}^{-1}$$

$$Y_p = \frac{QSC_{y_p}}{2mu_0} = \frac{6\,776.72 \times 0}{2 \times 85.4 \times 176} = 0$$

$$Y_r = 0$$

$$g/u_0 = (32.2/176)\text{ s}^{-1} = 0.183\text{ s}^{-1}$$

$$L_\beta = \frac{QSbC_{l_\beta}}{I_x} = \frac{6\,776.72 \times 33.4 \times (-0.074)}{1\,048}\text{ s}^{-2} = -15.98\text{ s}^{-2}$$

CHAPTER 5 Lateral Motion

$$L_p = \frac{QSb^2 C_{l_p}}{2I_x u_0} = \frac{6\ 776.72 \times 33.4^2 \times (-0.41)}{2 \times 1\ 048 \times 176}\ \text{s}^{-1} = -8.402\ \text{s}^{-1}$$

$$L_r = \frac{QSb^2 C_{l_r}}{2I_x u_0} = \frac{6\ 776.72 \times 33.4^2 \times 0.107}{2 \times 1\ 048 \times 176}\ \text{s}^{-1} = 2.193\ \text{s}^{-1}$$

$$N_\beta = \frac{QSb C_{n_\beta}}{I_z} = \frac{6\ 776.72 \times 33.4 \times 0.071}{3\ 530}\ \text{s}^{-2} = 4.552\ \text{s}^{-2}$$

$$N_p = \frac{QSb^2 C_{n_p}}{2I_z u_0} = \frac{6\ 776.72 \times 33.4^2 \times (-0.057\ 5)}{2 \times 3\ 530 \times 176}\ \text{s}^{-1} = -0.35\ \text{s}^{-1}$$

$$N_r = \frac{QSb^2 C_{n_r}}{2I_z u_0} = \frac{6\ 776.72 \times 33.4^2 \times (-0.125)}{2 \times 3\ 530 \times 176}\ \text{s}^{-1} = -0.760\ 5\ \text{s}^{-1}$$

$$\frac{Y_{\delta_r}}{u_0} = \frac{QS C_{y_{\delta_r}}/m}{u_0} = \frac{6\ 776.72 \times 0.157}{85.4 \times 176} = 0.070\ 8$$

$$L_{\delta_a} = \frac{QSb C_{l_{\delta_a}}}{I_x} = \frac{6\ 776.72 \times 33.4 \times (-0.134)}{1\ 048}\ \text{s}^{-2} = -28.941\ \text{s}^{-2}$$

$$L_{\delta_r} = \frac{QSb C_{l_{\delta_r}}}{I_x} = \frac{6\ 776.72 \times 33.4 \times 0.107}{1\ 048}\ \text{s}^{-2} = 23.11\ \text{s}^{-2}$$

$$N_{\delta_a} = \frac{QSb C_{n_{\delta_a}}}{I_z} = \frac{6\ 776.72 \times 33.4 \times (-0.003\ 5)}{3\ 530}\ \text{s}^{-2} = -0.224\ 4\ \text{s}^{-2}$$

$$N_{\delta_r} = \frac{QSb C_{n_{\delta_r}}}{I_z} = \frac{6\ 776.72 \times 33.4 \times (-0.072)}{3\ 530}\ \text{s}^{-2} = -4.617\ \text{s}^{-2}$$

Substituting the above values into the equation (5.1 - 3) yields the state equation as follows:

$$\begin{bmatrix} \Delta\dot\beta \\ \Delta\dot p \\ \Delta\dot r \\ \Delta\dot\phi \end{bmatrix} = \begin{bmatrix} -0.254\ 3 & 0 & -1 & 0.183 \\ -15.982 & -8.402 & 2.193 & 0.000 \\ 4.552 & -0.35 & -0.760\ 5 & 0.000 \\ 0.000 & 1 & 0.000 & 0.000 \end{bmatrix} \begin{bmatrix} \Delta\beta \\ \Delta p \\ \Delta r \\ \Delta\phi \end{bmatrix} + \begin{bmatrix} 0 & 0.070\ 8 \\ -28.941 & 23.11 \\ -0.224\ 4 & -4.617 \\ 0 & 0 \end{bmatrix} \begin{bmatrix} \Delta\delta_a \\ \Delta\delta_r \end{bmatrix}$$

MATLAB program 5.5 - 1 generates the coefficients of the characteristic polynomial and eigenvalues of the coefficient matrix.

5.5 Examples of Lateral Motion

MATLAB program 5.5 - 1

```
>> A=[-0.2543 0 -1 0.183;-15.982 -8.402 2.193 0;4.552 -0.35 -0.7605 0;0 1 0 0];
>> B=[0 0.0708;-28.941 23.11;-0.2244 -4.617;0 0];
>> C=[1 0 0 0;0 1 0 0;0 0 1 0;0 0 0 1];
>> D=[0 0;0 0;0 0;0 0];
>> poly(A)
ans =
    1.0000    9.4168   14.0393   48.5844    0.3974
>> eig(A)
>> eig(A)
ans =
   -8.4346
   -0.4870+2.3472i
   -0.4870-2.3472i
   -0.0082
```

Then the characteristic equation $|\lambda I - A| = 0$ is written as
$$\lambda^4 + 9.4168\lambda^3 + 14.0393\lambda^2 + 48.5844\lambda + 0.3974 = 0$$
The roots of the characteristic equation are
$$\lambda_{1,2} = -0.487 \pm i(2.3472) \quad \text{(Dutch roll mode)}$$
$$\lambda_3 = -8.4346 \quad \text{(Roll mode)}$$
$$\lambda_4 = -0.0082 \quad \text{(Spiral mode)}$$
The lateral modes are also estimated by means of the foregoing approximations.
The spiral mode characteristic value is
$$\lambda_{\text{spiral}} = \frac{L_\beta N_r - L_r N_\beta}{L_\beta} = \frac{(-15.98) \times (-0.7605) - 2.193 \times 4.552}{-15.98} \text{ s}^{-1} = -0.136 \text{ s}^{-1}$$
Compared with the accurate value, the spiral mode approximate value has some error.
The roll mode characteristic value is
$$\lambda_{\text{roll}} = L_p = -8.402 \text{ s}^{-1}$$
Compared with the accurate value, the roll mode approximate value is satisfactory.

Substituting the stability derivatives values into the Dutch roll motion approximate equation (5.3 - 11) yields
$$s^2 - \frac{-44.7549 + 176 \times (-0.7605)}{176} s + \frac{-44.7549 \times (-0.7605) - 4.552 \times 0 + 176 \times 4.552}{176} = 0$$
i.e.
$$s^2 + 1.019s + 4.745 = 0$$

MATLAB program 5.5 – 2 generates the roots of the above equation.

```
MATLAB program 5.5 – 2
>> p=[1 1.019 4.745];
>> roots(p)
ans =
   -0.5095+2.1179i
   -0.5095-2.1179i
```

That is
$$\lambda_{DR} = -0.5095 \pm i(2.1179)$$

Compared with the accuracy value, the Dutch roll mode approximate values are satisfactory.

MATLAB program 5.5 – 3 produce the zero-input responses of the system with the initial condition $\Delta\beta_0 = 2°$. The resulting responses are shown in Figure 5.5 – 1.

```
MATLAB program 5.5 – 3
>> x0=[2/57.3;0;0;0];
>> initial(A,B,C,D,x0,20);
```

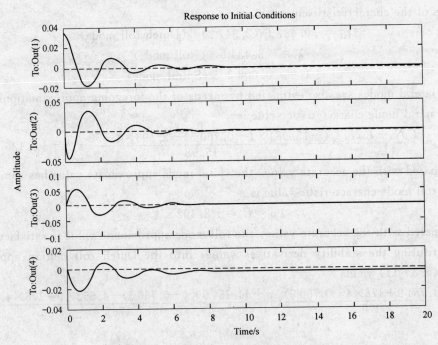

Figure 5.5 – 1 Zero-input responses

MATLAB program 5.5 – 4 generates the aileron impulse-input responses. The resulting responses are shown in Figure 5.5 – 2.

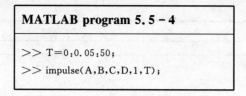

MATLAB program 5.5 – 4

\>\> T=0:0.05:50;
\>\> impulse(A,B,C,D,1,T);

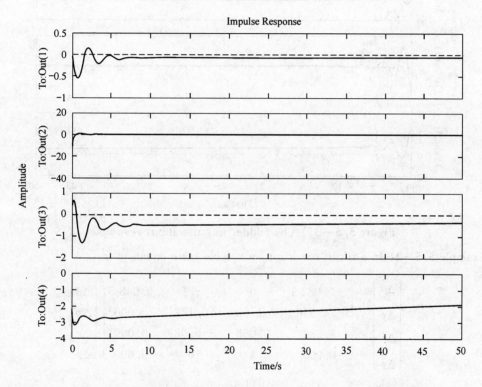

Figure 5.5 – 2 Aileron impulse-input response

MATLAB program 5.5 – 5 generates the responses to the rudder impulse-input. The resulting responses are shown in Figure 5.5 – 3.

MATLAB program 5.5 – 5

\>\> T=0:0.05:50;
\>\> impulse(A,B,C,D,2,T);

CHAPTER 5 Lateral Motion

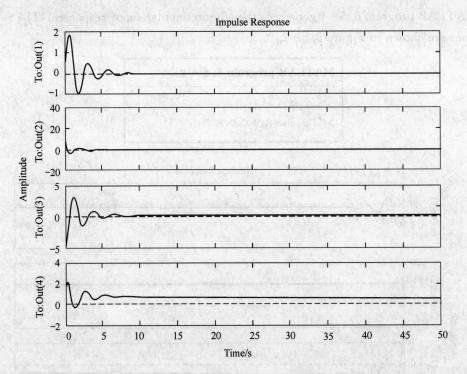

Figure 5.5 – 3 The rudder impulse-input response

Example 5.5 – 2: In a flight condition, a vehicle state equation is

$$\begin{bmatrix} \Delta\dot\beta \\ \Delta\dot p \\ \Delta\dot r \\ \Delta\dot\phi \end{bmatrix} = \begin{bmatrix} -0.0829 & 0 & -1 & 0.0485 \\ -4.546 & -1.699 & 0.172 & 0.000 \\ 3.382 & 0.065 & -0.089 & 0.000 \\ 0.000 & 1 & 0 & 0.000 \end{bmatrix} \begin{bmatrix} \Delta\beta \\ \Delta p \\ \Delta r \\ \Delta\phi \end{bmatrix} + \begin{bmatrix} 0 & 0.0116 \\ -27.276 & 0.576 \\ 0.395 & -1.362 \\ 0 & 0 \end{bmatrix} \begin{bmatrix} \Delta\delta_a \\ \Delta\delta_r \end{bmatrix}$$

MATLAB program 5.5 – 6 generates the coefficient of the characteristic polynomial and eigenvalues of the coefficient matrix.

5.5 Examples of Lateral Motion

MATLAB program 5.5-6

```
>> A=[-0.0829 0 -1 0.0485;-4.546 -1.699 0.172 0;3.382 0.065 -0.089 0;0 1 0 0];
>> B=[0 0.116;-27.276 0.576;0.395 -1.362;0 0];
>> C=[1 0 0 0;0 1 0 0;0 0 1 0;0 0 0 1];
>> D=[0 0;0 0;0 0;0 0];
>> poly(A)
ans =
   1.0000    1.8709    3.6703    5.6826   -0.0086
eig(A)
ans =
   -0.0911+1.8322i
   -0.0911-1.8322i
   -1.6903
    0.0015
```

Then the characteristic equation is written as
$$\lambda^4 + 1.8709\lambda^3 + 3.6703\lambda^2 + 5.6826\lambda - 0.0086 = 0$$

The roots of the characteristic equation are

$$\lambda_{1,2} = -0.0911 \pm i(1.8322) \quad \text{(Dutch roll mode)}$$
$$\lambda_3 = -1.6903 \quad \text{(Roll mode)}$$
$$\lambda_3 = 0.0015 \quad \text{(Spiral mode)}$$

Enter the following MATLAB commands to obtain the transfer functions.

```
>> [num,den]=ss2tf(A,B,C,D,1);
num =
    0   -0.0000   -0.3950   -0.2211   -0.1144
    0  -27.2760   -4.6208  -90.6474    0.0000
    0    0.3950   -1.0691   -0.0913   -4.3869
    0    0.0000  -27.2760   -4.6208  -90.6474
den =
    1.0000    1.8709    3.6703    5.6826   -0.0086
```

As an example, the transfer function $\Delta\beta(s)/\Delta\delta_a(s)$ is

$$\frac{\Delta\beta(s)}{\Delta\delta_a(s)} = -\frac{0.3953s^2 + 0.2211s + 0.1144}{s^4 + 1.8709s^3 + 3.6703s^2 + 5.6826s - 0.0086}$$

MATLAB program 5.5 – 7 generates the zero-input responses with the initial condition $\Delta\beta_0 = 2°$. The resulting responses are shown in Figure 5.5 – 4.

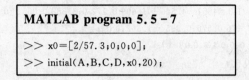

MATLAB program 5.5 – 7

\>\> x0=[2/57.3;0;0;0];
\>\> initial(A,B,C,D,x0,20);

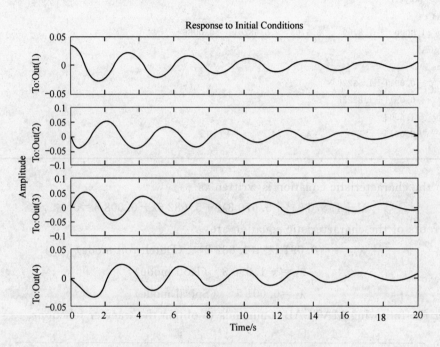

Figure 5.5 – 4 Zero-input response

MATLAB program 5.5 – 8 generates the aileron impulse-input responses. The resulting responses are shown in Figure 5.5 – 5.

MATLAB program 5.5 – 8

\>\> T=0:0.05:50;
\>\> impulse(A,B,C,D,1,T);

Enter the following MATLAB command
  \>\> impulse(A,B,C,D,2,T);
into the computer to obtain the rudder impulse-input responses. Figure 5.5 – 6 shows the responses to the rudder impulse-input.

5.5 Examples of Lateral Motion

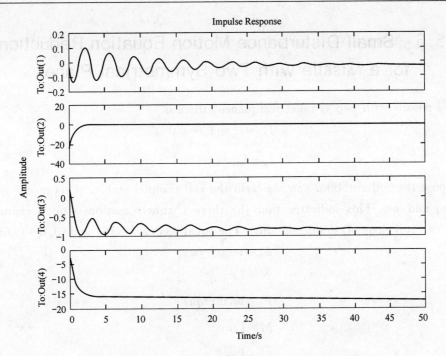

Figure 5.5 – 5 Aileron impulse-input responses

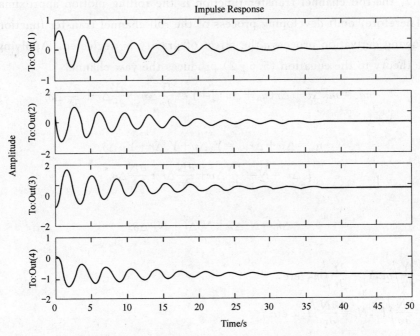

Figure 5.5 – 6 Rudder impulse-input responses

95

CHAPTER 5　Lateral Motion

5.6　Small Disturbance Motion Equation Reduction for a Missile with Two Symmetrical Planes

For a missile with two symmetrical planes, there is

$$\left.\begin{aligned} I_{xy}=I_{yz}=I_{zx}=0 \\ I_y=I_z \end{aligned}\right\} \quad (5.6-1)$$

Suppose the roll autopilot can maintain the roll channel stable. This means neglecting pv, pr, pq and pw. This indicates that the three channels motions (pitch channel, yaw channel, and roll channel) are decoupling. The equations (3.3-18) and (3.3-19) reduce to

$$\left.\begin{aligned} F_y &= m(\dot{v}+ru) \\ N &= I_z \dot{r} \end{aligned}\right\} \quad (5.6-2)$$

$$\left.\begin{aligned} F_z &= m(\dot{w}-qu) \\ M &= I_y \dot{q} \end{aligned}\right\} \quad (5.6-3)$$

$$L = I_x \dot{p} \quad (5.6-4)$$

Actually, the roll channel transfer function is the rolling motion approximate transfer function. Therefore, omit developing process of the roll channel transfer function.

In following derivation process, the effect of gravity is neglected. Applying the small disturbance theory to the equation (5.6-2) produces the yaw channel:

$$ma_y = m[\Delta\dot{v}+\Delta r(u_0+\Delta u)] = Y = \frac{\partial Y}{\partial v}\Delta v + \frac{\partial Y}{\partial r}\Delta r + \frac{\partial Y}{\partial \delta_r}\Delta\delta_r$$

i.e.

$$a_y = \Delta\dot{v}+\Delta r u_0 = Y_v \Delta v + Y_r \Delta r + Y_{\delta_r}\Delta\delta_r \quad (5.6-5)$$

$$I_z \Delta\dot{r} = N = \frac{\partial N}{\partial v}\Delta v + \frac{\partial N}{\partial r}\Delta r + \frac{\partial N}{\partial \delta_r}\Delta\delta_r$$

i.e.

$$\Delta\dot{r} = N_v \Delta v + N_r \Delta r + N_{\delta_r}\Delta\delta_r \quad (5.6-6)$$

where

$$Y_v = \frac{\partial Y}{\partial v}\bigg/m, \quad Y_r = \frac{\partial Y}{\partial r}\bigg/m, \text{ etc,}$$

$$N_v = \frac{\partial N}{\partial v}\bigg/I_z, \quad N_r = \frac{\partial N}{\partial r}\bigg/I_z, \text{ etc.}$$

In a similar manner, from the equation (5.6-3), the pitch channel is

5.6 Small Disturbance Motion Equation Reduction for a Missile with Two Symmetrical Planes

$$ma_z = m(\Delta\dot{w} - \Delta q u_0) = Z = \frac{\partial Z}{\partial w}\Delta w + \frac{\partial Z}{\partial q}\Delta q + \frac{\partial Z}{\partial \delta_e}\Delta\delta_e$$

i. e.

$$a_z = \Delta\dot{w} - \Delta q u_0 = Z_w \Delta w + Z_q \Delta q + Z_{\delta_e} \Delta\delta_e \tag{5.6-7}$$

$$I_y \Delta\dot{q} = M = \frac{\partial M}{\partial w}\Delta w + \frac{\partial M}{\partial q}\Delta q + \frac{\partial M}{\partial \delta_e}\Delta\delta_e$$

i. e.

$$\Delta\dot{q} = M_w \Delta w + M_q \Delta q + M_{\delta_e} \Delta\delta_e \tag{5.6-8}$$

where

$$Z_w = \frac{\partial Z}{\partial w}\bigg/m, \; Z_q = \frac{\partial Z}{\partial q}\bigg/m, \; \text{etc},$$

$$M_w = \frac{\partial M}{\partial w}\bigg/I_y, \; M_q = \frac{\partial M}{\partial q}\bigg/I_y, \; \text{etc}.$$

The equations (5.6-5) and (5.6-6) are rewritten in matrix form

$$\begin{bmatrix}\Delta\dot{v}\\\Delta\dot{r}\end{bmatrix} = \begin{bmatrix}Y_v & -(u_0 - Y_r)\\N_v & N_r\end{bmatrix}\begin{bmatrix}\Delta v\\\Delta r\end{bmatrix} + \begin{bmatrix}Y_{\delta_r}\\N_{\delta_r}\end{bmatrix}\Delta\delta_r \tag{5.6-9}$$

Taking Laplace transform of the equation (5.6-9) yields

$$(sI - A)X = B\Delta\delta_r$$

i. e.

$$\begin{bmatrix}s - Y_v & (u_0 - Y_r)\\-N_v & s - N_r\end{bmatrix}\begin{bmatrix}\Delta v\\\Delta r\end{bmatrix} = \begin{bmatrix}Y_{\delta_r}\\N_{\delta_r}\end{bmatrix}\Delta\delta_r \tag{5.6-10}$$

Use Cramer's rule to obtain the transfer functions as follows:

$$\frac{\Delta v(s)}{\Delta\delta_r(s)} = \frac{\begin{vmatrix}Y_{\delta_e} & u_0 - Y_r\\N_{\delta_r} & s - N_r\end{vmatrix}}{\begin{vmatrix}s - Y_v & (u_0 - Y_r)\\-N_v & s - N_r\end{vmatrix}} = \frac{Y_{\delta_r}s - N_{\delta_r}u_0 - Y_{\delta_r}N_r + N_{\delta_r}Y_r}{s^2 - (Y_v + N_r)s + Y_v N_r + u_0 N_v - N_v Y_r} \tag{5.6-11}$$

$$\frac{\Delta r(s)}{\Delta\delta_r(s)} = \frac{\begin{vmatrix}s - Y_v & Y_{\delta_e}\\-N_v & N_{\delta_r}\end{vmatrix}}{\begin{vmatrix}s - Y_v & (u_0 - Y_r)\\-N_v & s - N_r\end{vmatrix}} = \frac{N_{\delta_r}s - N_{\delta_r}Y_v + N_v Y_{\delta_r}}{s^2 - (Y_v + N_r)s + Y_v N_r + u_0 N_v - N_v Y_r} \tag{5.6-12}$$

CHAPTER 5 Lateral Motion

If Y_r is very small and neglected, then

$$\frac{\Delta v(s)}{\Delta \delta_r(s)} = \frac{Y_{\delta_r} s - N_{\delta_r} u_0 - Y_{\delta_r} N_r}{s^2 - (Y_v + N_r)s + Y_v N_r + u_0 N_v} \qquad (5.6-13)$$

$$\frac{\Delta r(s)}{\Delta \delta_r(s)} = \frac{N_{\delta_r} s - N_{\delta_r} Y_v + N_v Y_{\delta_r}}{s^2 - (Y_v + N_r)s + Y_v N_r + u_0 N_v} \qquad (5.6-14)$$

Because of $\Delta \beta = \Delta v / u_0$,

$$\frac{\Delta \beta(s)}{\Delta \delta_r(s)} = \frac{(Y_{\delta_r} s - N_{\delta_r} u_0 - Y_{\delta_r} N_r)/u_0}{s^2 - (Y_v + N_r)s + Y_v N_r + u_0 N_v} \qquad (5.6-15)$$

Considering $a_y = \Delta \dot{v} + \Delta r u_0$,

$$\frac{a_y(s)}{\Delta \delta_r(s)} = \frac{Y_{\delta_r} s^2 - Y_{\delta_r} N_r s - u_0 (N_{\delta_r} Y_v - N_v Y_{\delta_r})}{s^2 - (Y_v + N_r)s + Y_v N_r + u_0 N_v} \qquad (5.6-16)$$

The equations (5.6-7) and (5.6-8) are rewritten in matrix form

$$\begin{bmatrix} \Delta \dot{w} \\ \Delta \dot{q} \end{bmatrix} = \begin{bmatrix} Z_w & u_0 + Z_q \\ M_w & M_q \end{bmatrix} \begin{bmatrix} \Delta w \\ \Delta q \end{bmatrix} + \begin{bmatrix} Z_{\delta_e} \\ M_{\delta_e} \end{bmatrix} \delta_e \qquad (5.6-17)$$

Taking Laplace transform of the equation (5.6-17) yields

$$(sI - A)X = B\delta_e$$

i.e.

$$\begin{bmatrix} s - Z_w & -(u_0 + Z_q) \\ -M_w & s - M_q \end{bmatrix} \begin{bmatrix} \Delta w \\ \Delta q \end{bmatrix} = \begin{bmatrix} Z_{\delta_e} \\ M_{\delta_e} \end{bmatrix} \delta_e$$

Using Cramer's rule, one obtains the transfer function as follows:

$$\frac{\Delta w(s)}{\Delta \delta_e(s)} = \frac{\begin{vmatrix} Z_{\delta_e} & -(u_0 + Z_q) \\ M_{\delta_e} & s - N_r \end{vmatrix}}{\begin{vmatrix} s - Z_w & -(u_0 + Z_q) \\ -M_w & s - M_q \end{vmatrix}} = \frac{Z_{\delta_e} s + M_{\delta_e} u_0 - Z_{\delta_e} M_q}{s^2 - (Z_w + M_q)s + Z_w M_q - u_0 M_w - M_w Z_q}$$

$$(5.6-18)$$

$$\frac{\Delta q(s)}{\Delta \delta_e(s)} = \frac{\begin{vmatrix} s - Z_w & Z_{\delta_e} \\ -M_w & M_{\delta_e} \end{vmatrix}}{\begin{vmatrix} s - Z_w & -(u_0 + Z_q) \\ -M_w & s - M_q \end{vmatrix}} = \frac{M_{\delta_e} s - M_{\delta_e} Z_w + Z_{\delta_e} M_w}{s^2 - (Z_w + M_q)s + Z_w M_q - u_0 M_w - M_w Z_q}$$

$$(5.6-19)$$

5.6 Small Disturbance Motion Equation Reduction for a Missile with Two Symmetrical Planes

If Z_q is very small and neglected, then

$$\frac{\Delta w(s)}{\Delta \delta_e(s)} = \frac{Z_{\delta_e} s + M_{\delta_e} u_0 - Z_{\delta_e} M_q}{s^2 - (Z_w + M_q)s + Z_w M_q - u_0 M_w} \tag{5.6-20}$$

$$\frac{\Delta q(s)}{\Delta \delta_e(s)} = \frac{M_{\delta_e} s - M_{\delta_e} Z_w + Z_{\delta_e} M_w}{s^2 - (Z_w + M_q)s + Z_w M_q - u_0 M_w} \tag{5.6-21}$$

Considering $\Delta \alpha = \Delta w / u_0$,

$$\frac{\Delta \alpha(s)}{\Delta \delta_e(s)} = \frac{(Z_{\delta_e} s + M_{\delta_e} u_0 - Z_{\delta_e} M_q)/u_0}{s^2 - (Z_w + M_q)s + Z_w M_q - u_0 M_w} \tag{5.6-22}$$

In view of $a_z = \Delta \dot{w} - \Delta q u_0$, then

$$\frac{a_z(s)}{\Delta \delta_e(s)} = \frac{Z_{\delta_e} s^2 - Z_{\delta_e} M_q s - u_0(-M_{\delta_e} Z_w + M_w Z_{\delta_e})}{s^2 - (Z_w + M_q)s + Z_w M_q - u_0 M_w} \tag{5.6-23}$$

As an example, consider a typical surface-to-air missile with rear controls. The yaw derivatives for $Ma = 1.4$ ($u_0 = 467$ m/s) and height 1 500 m are

$$Y_v = -2.74, \quad N_v = 0.309, \quad Y_{\delta_r} = 197, \quad N_{\delta_r} = -534, \quad N_r = -2.89$$

The missile is 2 m long. $I_z = 13.8$ kg·m^2, $m = 53$ kg.
Substituting above data into equation (5.6-16) yields

$$\frac{a_y(s)}{\Delta \delta_r(s)} = \frac{197s^2 + 570s - 467(1\,460 - 60.8)}{s^2 + (2.74 + 2.89)s + (7.9 + 144)}$$

The natural frequency is

$$\omega^2 = 7.9 + 144 = 152, \quad \omega = 12.4 \text{ rad/s}$$

The damping ratio is

$$\mu = \frac{2.74 + 2.89}{2 \times 12.4} = 0.23$$

The steady state gain is

$$K_a = \frac{-467(1\,460 - 60.8)}{7.9 + 144} = -4\,300$$

MATLAB program 5.6-1 gives the step response curve of the system. The step response curve is shown in Figure 5.6-1.

MATLAB program 5.6-1
`>> num=[197 570 -467*(1460-60.8)];`
`>> den=[1 (2.74+2.89) (7.9+144)];`
`>> T=0:0.02:1.5;`
`>> step(num,den,T);`

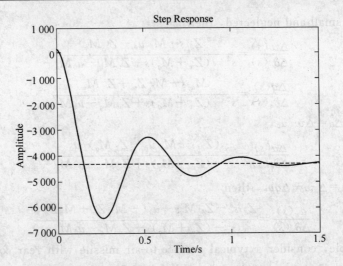

Figure 5.6-1 Step response for rear controls

Now consider the $\Delta\beta$ and Δr dynamic behaviors.

Substituting the given data into the equation (5.6-9) yields

$$\begin{bmatrix} \Delta\dot{v} \\ \Delta\dot{r} \end{bmatrix} = \begin{bmatrix} -2.74 & -467 \\ 0.309 & -2.89 \end{bmatrix} \begin{bmatrix} \Delta v \\ \Delta r \end{bmatrix} + \begin{bmatrix} 197 \\ -534 \end{bmatrix} \delta_r$$

Considering $\beta = \Delta v / u_0$,

$$\begin{bmatrix} \Delta\dot{\beta} \\ \Delta\dot{r} \end{bmatrix} = \begin{bmatrix} -2.74 & -1 \\ 0.309 \times 467 & -2.89 \end{bmatrix} \begin{bmatrix} \Delta\beta \\ \Delta r \end{bmatrix} + \begin{bmatrix} \dfrac{197}{467} \\ -534 \end{bmatrix} \delta_r$$

MATLAB program 5.6-2 generates the zero-input response curve of the system with $\Delta\beta_0 = 1°$. The zero-input response curve is shown in Figure 5.6-2.

MATLAB program 5.6-2

```
>> A=[-2.74 -1;0.309*467 -2.89];
>> B=[197/467;-534];
>> C=[1 0;0 1];
>> D=[0;0];
>> eig(A)
ans =
    -2.8150+12.0124i
    -2.8150-12.0124i
>> x0=[1/57.3;0];
>> initial(A,B,C,D,x0,1.5);
```

5.6 Small Disturbance Motion Equation Reduction for a Missile with Two Symmetrical Planes

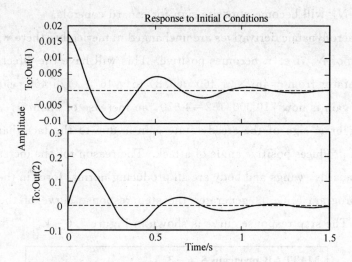

Figure 5.6 – 2   Zero-input response for rear controls

The step response curves are obtained through entering the following MATLAB commands
>> T=0:0.01:1.8;
>> step(A,B,C,D,1,T);

into the computer. The resulting step response curves are shown in Figure 5.6 – 3.

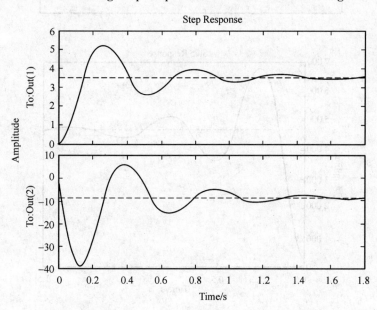

Figure 5.6 – 3   Step response for rear controls

CHAPTER 5　Lateral Motion

But M_{δ_e} and N_{δ_e} will become opposite sign for canard controls.

Assume all aerodynamic derivatives are unchanged numerically, there will be a change in the algebraic sign of N_{δ_r}, i.e. it becomes positive. This will have the effect of changing one term in the numerator from $-467 \times (1\,460 - 60.8)$ to $-467 \times (-1\,460 - 60.8) = 710\,000$. The steady state gain is now $710\,000/152 = 4\,670$, an increase of about 9%. The reason for the change in algebraic sign of the steady state gain is due to the fact that positive canard rudder deflection produces positive angle of attack. The reason for the increase in gain is due to the fact that canards, wings and body are all producing normal force in the same direction.

MATLAB program 5.6 - 3 generates the step response curve of the system for the canard controls. The step response curve is shown in Figure 5.6 - 4.

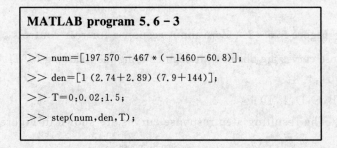

MATLAB program 5.6 - 3

```
>> num=[197 570 -467*(-1460-60.8)];
>> den=[1 (2.74+2.89) (7.9+144)];
>> T=0:0.02:1.5;
>> step(num,den,T);
```

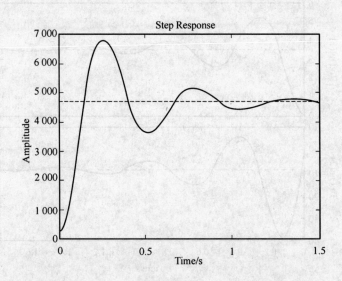

Figure 5.6 - 4　Step response for canard controls

5.6 Small Disturbance Motion Equation Reduction for a Missile with Two Symmetrical Planes

References

[1] Robert C Nelson. Flight Stability and Automatic Control. New York:McGraw-Hill, 1989.
[2] Etkin B. Dynamics of Flight Stability and Control. New York:John Wiley, 1996.
[3] John H Blakelock. Automatic Control of Aircraft and Missiles. 2nd ed. New York:John Wiley, 1991.
[4] Garnell P. Guided Weapon Control Systems. Oxford:Pergamon Press,1980.

CHAPTER 6
Flight Control of Missile

6.1 Introduction

In order to finish a flight mission, the flight velocity of a missile needs to be controlled. The flight velocity change depends on the external force acting on the missile. When a missile is flying in the air, the thrust, the aerodynamic force, and the gravity act on the missile. However, only the thrust T and the aerodynamic force R may be changed. If N denotes the resultant force of the thrust and the aerodynamic force, then

$$N = T + R \qquad (6.1-1)$$

The resultant force N is resolved into a tangent force N_t and a normal force N_n (as shown in Figure 6.1-1), i.e.

$$N = N_t + N_n \qquad (6.1-2)$$

The tangent force N_t can change the magnitude of flight velocity, while the normal force can change the direction of flight velocity. In general, for missile, the magnitude of flight velocity is uncontrolled.

For winged missiles and wingless missiles, the methods of generating and changing force are different. The winged missiles are usually

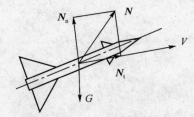

Figure 6.1-1 Tangent force N_t and normal force N_n

controlled by the aerodynamic force since they fly in dense atmosphere, whereas the wingless missiles are usually controlled by the engine thrust since they fly in rarefied air or outside the atmosphere.

For most guided missiles, the direction of the flight velocity can be changed in light of guidance commands. In general, for a winged missile, to change the normal force magnitude, it is necessary to rotate the missile an angle so as to change attack angle α, sideslip angle β, and roll angle ϕ. In order to rotate the missile, it is necessary to apply properly a control moment relative to the center of mass on the missile. The element generating the moment is called control element, or called control surface, such as aerodynamic fin. It can create aerodynamic force Z_δ due to deflection. The moment produced by Z_δ about the center of mass is the control moment M_δ.

In addition, control surfaces also play the role of stabilization during flight. When a vehicle is flying, the vehicle encounters disturbances such as atmospheric disturbance, so as to depart from original equilibrium state. In this circumstance, deflecting control surfaces can cause the vehicle to go back to the original equilibrium state. Thus, control surfaces are used for control and stabilization.

6.2 Control Force Generation

6.2.1 Aerodynamic Force Control

Aerodynamic force method is a typical method. Traditionally, control surfaces generate control moments to rotate the missile about the center of mass. The result changes angle of attack. This means the lift change occurs. The lift is perpendicular to the flight velocity. The side force is also perpendicular to the flight velocity. But the drag is opposite from the flight direction. Therefore, controlling normal forces actually change the lift and side force. Considering some missiles with a symmetric plane and the others with two symmetric planes, they will be discussed respectively.

1. Missile with two symmetric planes

For this type of missiles, there are two pairs of fins which can generate aerodynamic control forces in two cross planes. Figure 6.2 – 1 shows the cruciform configuration of missiles.

For the cruciform configuration missiles, they are controlled, usually using STT (skid-to-turn). For "+" type missiles, if deflecting 1, 3 control surfaces upwards (or downwards) generates control positive (or negative) pitching moment along Oy_b axis, then the missile's nose pitches up (down) to yield an angle of attack, α. The lift changes with the variation of α.

CHAPTER 6　Flight Control of Missile

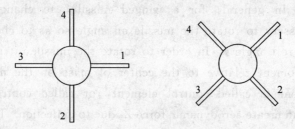

Figure 6.2 – 1　Control surfaces looking from the rear of missile

The 1, 3 control surfaces act as elevators. If deflecting 2, 4 control surfaces leftward (or rightward) generates negative (or positive) yawing control moment along Oz_b axis, then the missile's nose rotates leftward (or rightward) to yield an sideslip angle, β. The side force changes with the variation of β. The 2, 4 control surfaces play the role of rudders.

For "x" type missiles, if deflecting 1, 3 control surfaces and 2, 4 control surfaces downward (or upward) through the same angle (as indicated in Figure 6.2 – 2 (a)) generates pitching control moment, then two pairs of fins 1, 3, 2, 4 act as elevators. If deflecting two pairs of control surfaces inversely through the same angle (as shown in Figure 6.2 – 2 (b)) generates yawing control moment, then two pairs of fins 1, 3, 2, 4 act as rudders. Deflecting two pairs of control surfaces at different angles (as shown in Figure 6.2 – 2 (c)) yields a resultant steering force. The two pairs of control surfaces act as elevators and rudders.

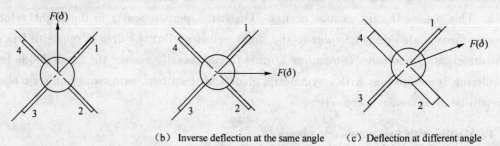

(b) Inverse deflection at the same angle　　(c) Deflection at different angle

Figure 6.2 – 2　Two pairs of fins are deflected for different angles

If anyone of a pair of control surfaces has its own actuator, the control surfaces are deflected not only in the same direction, but also differentially. The control surfaces of differential deflection play the role of ailerons.

For the cruciform configuration missiles, the function of roll channel is to provide missile stabilization of roll attitude about the longitudinal axis. In other word, the rolling is not used to turn. Therefore, the pitching channel, yawing channel, and rolling channel are considered to be independent, and the couplings are neglected. When the missile yaws in horizontal plane, the forces and moments causing yawing motion are similar to those resulting in pitching motion.

For the cruciform configuration missiles, rear control and canard control configuration are widely used. For the rear control configuration (called normal configuration), the wings are in front of the control surfaces. The wings supply the lift and side force. Deflecting the fins upward (or downward) causes the angle of attack change. But the lift increment on the fins is opposite from the lift increment on the wings. Thus, resultant lift decreases slightly. For the canard configuration, the control surfaces are in front of the wings. Deflecting the control surfaces causes the angle of attack change. The lift increment on the fins is the same as the lift increment on the wings. Therefore, resultant lift increases slightly. In view of the point, the canard control efficiency is higher than that of the rear control.

2. Missile with a symmetric plane

The missiles configuration is like airplane as seen in Figure 6.2 - 3. The longitudinal channel control is similar to the cruciform missile. For this type of missile, BTT (bank-to-turn) is widely used for lateral-directional channels. The horizontal component of the lift (see Figure 6.2 - 3) is side force. To turn in horizontal plane, the missile not only has a roll angle, but also properly increases angle of attack to ensure that the vertical component of the lift can balance the gravity. Otherwise, the missile loses height.

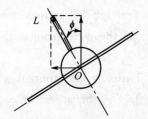

Figure 6.2 - 3 Bank-to-turn

In light of requirement of sideslip angle, there are coordination control and non-coordination control. The former retains the sideslip angle to be approximately zero while the latter does not so. There is coupling between the roll channel and yaw channel in BTT.

6.2.2 Thrust Vector Control

So called thrust vector control is a method of steering a missile. That is to operate the missile by altering the direction of the efflux from the propulsion motor. Evidently, the control method is not essentially dependant on the dynamic pressure of the atmosphere. The thrust vector can perform effectively whether in rare atmosphere, outside atmosphere, or at low speed. Clearly, it is inoperative after motor burn-out. For example, Figure 6.2 - 4 indicates the forces acting on the missile if the thrust T is deflected through angle δ.

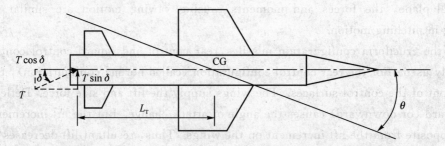

Figure 6.2 - 4 Forces due to thrust vector deflection

If the mass of the missile is 30 kg, the moment of inertia about the center of gravity is 1 kg·m² (both assumed constant), the thrust is 1 500 N and the moment arm l_T is 0.35 m, then for $\delta = 4°$,

$$\ddot{\theta} = -\frac{T \sin \delta l_T}{J} = \frac{1\,500 \times (\sin 4°) 0.35}{1} \text{ rad/s}^2 = -36.75 \text{ rad/s}^2 \qquad (6.2-1)$$

If requiring the missile to rotate 90° to rest, the ideal control system will accelerate the body through 45° at maximum acceleration and then use full braking moment for the same time. In these conditions, the time for a complete turn from rest to rest will be taken about 0.4 s.

Thrust vector control is usually used for the following situations:

① The thrust vector control is applied in the vertical launch phase of ballistic missile. Since a ballistic missile is heavy in launch phase, the velocity is slow. Aerodynamic force control is ineffective. To prevent from toppling over due to a small inevitable thrust misalignment, it is necessary to use the thrust vector control.

② The thrust vector control is applied in the vertical launch phase of tactical missiles. To intercept targets in all directions, by using the thrust vector control, the missile turns as soon as possible after launch. In addition, vertical launch reduces launcher and saves launch space (it is especially useful in ships).

③ For some short range missiles, such as anti-tank missiles, in order to maneuver immediately after launch, it is effective to use the thrust vector control.

The implementing methods of thrust vector in general have ① gimbaled motors, ② ball and socket or flexible nozzles, ③ spoilers or vanes, and ④ secondary fluid or gas injection.

6.2.3 Rocket Injection Control

1. Elbow nozzle deflection control

Figure 6.2−5(a) shows that the elbow nozzle steers missile.

The elbow nozzles are set around the center of gravity. The efflux from the nozzle generates the counterforce. The direction of the force varies with the nozzle deflection. Referring to Figure 6.2−5(a), then the normal force N_n is

$$N_n = T\sin\delta \qquad (6.2-2)$$

where

T = thrust resulting from injection.

2. Side nozzle direct control

Some rocket nozzles are vertically set around the center of mass. The axes of nozzles are perpendicular to the missile's longitudinal axis. The distribution device guides the efflux from a rocket engine to nozzles (as seen in Figure 6.2−5(b)). Except the manner of single rocket engine for the nozzles, multi-engines are also used. Each nozzle corresponds to a small engine. For instance, in American PAC—3 missile, 180 pulse solid fuel rocket engines are installed around the body. 10 engines of a column are uniformly distributed at 20° intervals around the body. Each engine generates 2 500 N thrust in 0.02 seconds.

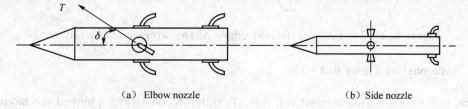

(a) Elbow nozzle   (b) Side nozzle

Figure 6.2−5 Steer missile relying on counterforce

6.3 Steering Components

6.3.1 Aerodynamic Control Surfaces

The aerodynamic control surfaces, which are set at a distance from the center of mass, are widely used for steering missile.

Aerodynamic control surfaces usually include types below.

1. All-moving fins (see Figure 6.3-1)

Entire fin may move. The fins are widely applied to supersonic missiles.

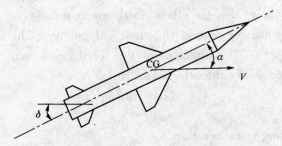

Figure 6.3-1　All-moving fins

2. Rear edge fins behind the wings and the tail wings (see Figure 6.3-2)

They are widely used for missiles like airplanes.

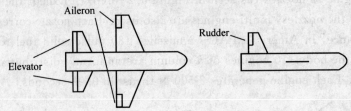

Figure 6.3-2　Fins locate rear edges of the wings and the tail wings

3. Rollerons (see Figure 6.3-3)

The rollerons are used to limit roll rate. A rolleron consists of a hinged tab mounted on the outer trailing edge of each wing panel. The tab contains a toothed wheel which is driven at a high angular velocity by the impinging air-stream.

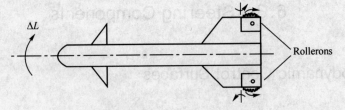

Figure 6.3-3　Rollerons

A rolleron is regarded as a one degree of freedom gyro. If the missile has a roll rate, the tab is caused to precess into the airstream. The resulting moment opposes the missile roll rate. If the hinge line is canted at 45° to the missile transverse axis, some damping in pitch and yaw is also achieved.

Compared with common aileron, the rollerons does not need actuators. Moreover, the structure of the rolleron is simple. They are often employed in mid-small type of missiles.

6.3.2 Jet Steering Components

1. Jet vanes

Jet vans are installed at engine nozzle exit (as seen in Figure 6.3 - 4). They are made of fire-retardant material. Like aerodynamic fins the jet vanes are deflected in jet stream to steer the missile pitch motion, yaw motion, and roll motion. The structure is simple; moreover they can work without the air. The engine thrust loss results from the resistance of the jet vanes. Of course there is a severe erosion problem because they are permanently in the very hot jet stream. The moving vanes were used in the German V-2 missile in World War II.

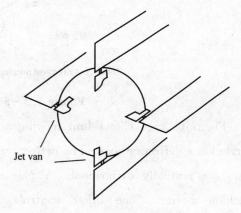

Figure 6.3 - 4 Jet vans

2. Gimbaled engines

The combustion chambers of liquid engines are fixed in gimbal (as seen in Figure 6.3 - 5). The combustion chambers may be deflected to change hot gas stream direction. A pair of combustion chambers deflected differentially may control roll motion of missile.

3. Moving nozzles

For solid engines, the combustion and fuel commonly share a chamber, so it is inconvenient to directly deflect the chamber. Moving nozzles are usually used. The joints between the combustion chamber and the nozzle include two means: ① flexible rubber mounting, and ② ball and socket joint. By deflecting the nozzle of engine, the jet direction can be changed. This results in the change of thrust direction.

CHAPTER 6 Flight Control of Missile

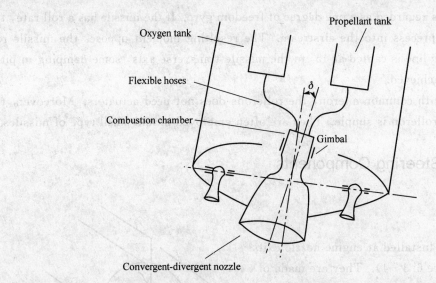

Figure 6.3 – 5 Gimbaled engines

Figure 6.3 – 6 shows four moving nozzles. Each nozzle has a deflection axis. The deflection axes of two pairs are mutually orthogonal. A pair controls the pitching motion. The other controls the yawing motion. Differential deflection of arbitrary one pair controls the rolling motion.

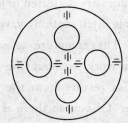

Figure 6.3 – 6 Four moving nozzles

6.4 Missile Maneuverability and Load Factor

The maneuverability is the very important one of a missile capability. The maneuverability can be obtained through changing both magnitude and direction of flight velocity. The maneuverability is evaluated with the tangent acceleration and normal acceleration. The tangent acceleration represents the capability of altering the magnitude of velocity. The normal acceleration expresses turning capability. For missiles, people are interested in normal acceleration, i.e. normal maneuverability. The better the maneuverability, the smaller the turning radius will be. The load factor may be used to evaluate the maneuverability of missile.

6.4 Missile Maneuverability and Load Factor

The load factor n is ratio of the resultant force N except gravity to the gravity G, i.e.

$$n = \frac{N}{G} \tag{6.4-1}$$

or

$$n = \frac{N}{mg} = \frac{N/m}{g} = \frac{a}{g} \tag{6.4-2}$$

The load factor is dimensionless quantity, and represents relative acceleration in unit g. It is a vector, and its direction coincides with the resultant force N.

The load factor may be projected in body system, wind axis system, and path system, etc. The projection n_{x_k} along the Ox_k in the flight path system is termed tangent load factor. The projection n_{y_k} along the Oy_k in the path system is termed lateral load factor. The projection n_{z_k} along the Oz_k in the path system is termed normal load factor. Sometimes the lateral load factor and normal load factor are called by a joint name of normal load factor.

If n_x denotes a component of the load factor n along the Ox axis in the body system, n_y denotes a component of the load factor n along the Oy axis in the body system, and n_z denotes a component of the load factor n along the Oz axis in the body system, then

$$\left. \begin{array}{l} n_x = \dfrac{N_x}{G} = \dfrac{a_x}{g} \\[4pt] n_y = \dfrac{N_y}{G} = \dfrac{a_y}{g} \\[4pt] n_z = \dfrac{N_z}{G} = \dfrac{a_z}{g} \end{array} \right\} \tag{6.4-3}$$

$$n = n_x \boldsymbol{i} + n_y \boldsymbol{j} + n_z \boldsymbol{k} \tag{6.4-4}$$

When discussing missile guidance, the relative motion relation between a missile and a target needs to be considered. Figure 6.4-1 indicates the relative motion relation. Supposing the missile and the target are flying in same plane, the load factor n_{LOS} along line of sight MT and the load factor $n_{\perp \text{LOS}}$ perpendicular to line of sight MT are respectively

$$n_{\text{LOS}} = n_x \cos(\theta - \lambda) - n_z \sin(\theta - \lambda) \tag{6.4-5}$$

$$n_{\perp \text{LOS}} = n_x \sin(\theta - \lambda) + n_z \cos(\theta - \lambda) \tag{6.4-6}$$

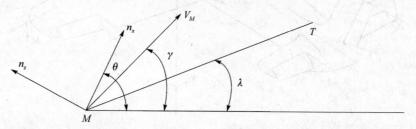

Figure 6.4-1 Relative motion relation between a missile and a target

Required load factor n_r means that a missile needs load factor along a given kinematical given trajectory. During practical flight, the missile can not completely fly along kinematical trajectory due to inertia and disturbances, so this means to add the load factor. Thus, available load factor n_a should be greater than the required load factor, i.e.

$$n_r < n_a \qquad (6.4-7)$$

Evidently, for a missile, the available load factor is expected as large as possible. Actually, the load factor is restricted by the following: ① steering component deflection limit, ② angles of attack and sideslip limit, and ③ structural strength limit.

6.5 Control Surface Specification

For the missile like airplane, there are steering elements including aileron, rudder, and elevator. For normal configuration, positive deflection of a control surface produces a negative steering moment.

Specify that deflecting elevator δ_e downward is positive. See Figure 6.5-1. The positive δ_e generates negative pitching moment M. Then the nose of missile will go downward.

Specify that deflecting right aileron downward (at the same time left aileron upward) is positive. See Figure 6.5-1. Positive deflection δ_a generates negative rolling moment. Then the right wing will go upward; the left wing will go downward.

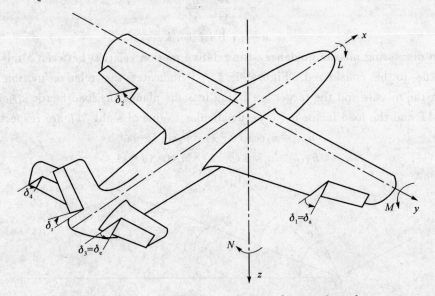

Figure 6.5-1 Deflecting polarity of control surface

6.5 Control Surface Specification

Referring to Figure 6.5 – 1, deflecting rudder δ_r leftward is specified to be positive. The positive δ_r produces negative yawing moment N. Then the nose of missile will go leftward.

It is noted that deflections along body axes are positive. Then

$$\delta_a = \frac{1}{2}(\delta_1 - \delta_2)$$

If $\delta_1 = -\delta_2$, then $\delta_a = \delta_1$.

For elevator,

$$\delta_e = \delta_3 = \delta_4$$

For cruciform missile, there is distribution about control surfaces in Figure 6.5 – 2.

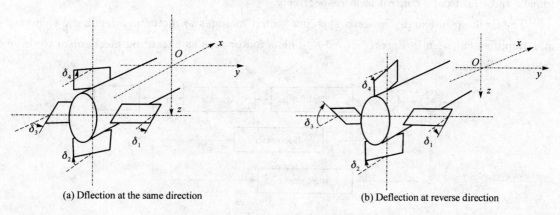

(a) Dflection at the same direction (b) Deflection at reverse direction

Figure 6.5 – 2 Control surfaces of missile

If deflections along body axes are positive, for elevator

$$\delta_e = 0.5(\delta_1 + \delta_3)$$

If $\delta_1 = \delta_3$, then $\delta_e = \delta_1 = \delta_3$.

For rudder δ_r,

$$\delta_r = 0.5(\delta_2 + \delta_4)$$

If $\delta_2 = \delta_4$, then $\delta_r = \delta_2 = \delta_4$.

For aileron δ_a,

$$\delta_a = \frac{1}{4}(\delta_1 - \delta_3 + \delta_2 - \delta_4)$$

If only two surfaces act differentially,

$$\delta_a = 0.5(\delta_1 - \delta_3) \quad \text{or} \quad \delta_a = 0.5(\delta_2 - \delta_4)$$

6.6 Flight Control System with Attitude Control

6.6.1 Control System Components

For the missile like airplane, such as cruise missile, the attitude control is universally used. The attitude control loop acts as inner loop. A typical automatic flight control system usually consists of three control loops, which are fin loop, stabilizing loop (attitude control loop), and flight path control loop respectively.

The fin loop is usually a servo system, which includes an actuator, feedback elements and amplifier (as seen in Figure 6.6 - 1). The actuator acts as executing mechanism to drive the fin deflection.

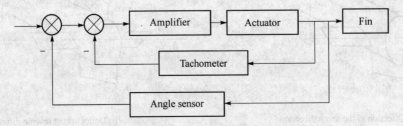

Figure 6.6 - 1 Fin loop

The angle feedback ensures corresponding relation between the input deflection command and control surface deflection. The angular speed feedback increases the system damping to improve dynamic performance of the fin loop.

If the sensor measures the attitude of vehicle, then the sensor, controller, and fin loop form autopilot. The autopilot and the vehicle compose stabilizing loop as shown in Figure 6.6 - 2. This loop stabilizes the vehicle's attitude, i.e. the angular motion of vehicle.

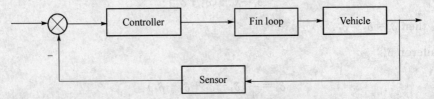

Figure 6.6 - 2 Stabilizing loop

The stabilizing loop, a device measuring position of the mass center, and kinematical element representing space geometrical relation form control loop as shown in Figure 6.6-3.

The Figure 6.6-3 is a block diagram of altitude control. The pitch angle control loop acts as inner loop of altitude loop. The altitude sensor feedback forms a trajectory control loop. The function of the altitude control is to hold given altitude flight.

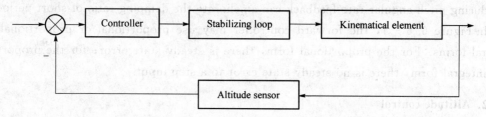

Figure 6.6-3 Control loop

6.6.2 Longitudinal Control

The perturbation motion of a vehicle usually consists of long-period mode and short-period mode. As flight envelop extends, in order to make the vehicle satisfy performance requirements in entire envelop, feedback control techniques are always used. For example, if a vehicle is flying at high altitude, the damping characteristic becomes bad, and the performance of short-period tends to deteriorate, aggravating oscillation. These severely affect the completion of the flight mission. Thus, introducing appropriate feedbacks in longitudinal channel may improve flight quality.

1. Longitudinal inner loop control

Longitudinal inner loop usually consists of pitch angle feedback and pitch angular rate feedback. Control structure is indicated in Figure 6.6-4.

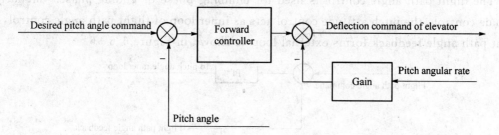

Figure 6.6-4 Block diagram of pitch angle control

The pitch angle feedback improves long-period mode performance such as increasing long-period damping (increasing long-period mode stability). The angular rate of pitch feedback improves short-period mode performance. Pitch angle feedback can increase long-period damping greatly, but at the same time decreases short-period damping. Thus, introducing pitch angular rate feedback can ameliorate the damping ratio of short-period.

In Figure 6.6 – 4, the forward controller may use proportional or proportional plus integral forms. For the proportional form, there is steady state error; for the proportional plus integral form, there is no steady state error to a step input.

2. Altitude control

The altitude control consists of pitch inner loop and external loop. The inner loop control is the same with the above inner control. Altimeter is a sensing device. The external loop of altitude control commonly uses proportional plus derivative form as shown in Figure 6.6 – 5.

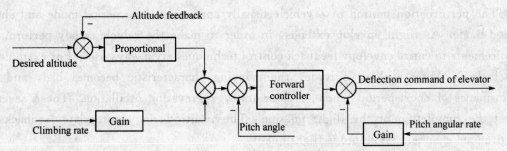

Figure 6.6 – 5 Block diagram of altitude control

3. Longitudinal flight path angle control

The flight path angle control is used for climbing phase or gliding phase. Similarly to altitude control, the pitch attitude control acts as inner loop of flight path angle control, and flight path angle feedback forms external loop as shown in Figure 6.6 – 6.

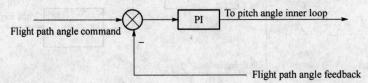

Figure 6.6 – 6 Block diagram of flight path angle control

6.6.3 Lateral Directional Control

For the missile with two symmetrical planes, the lateral channel and roll channel are commonly separated to be designed. In the case, the roll channel keeps the wings level. The lateral channel is similar to the longitudinal channel. The roll channel is indicated in Figure 6.6 – 7. The roll controller may be proportional or proportional plus integral forms. For a large natural damping, the angular velocity feedback may be omitted.

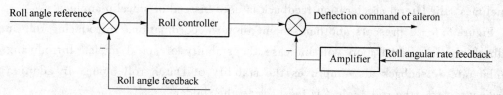

Figure 6.6 – 7 Block diagram of roll angle control

For the missile like airplane, which is with a symmetrical plane, a coordinated turn is used. The method of coordination attempts to keep that the longitudinal axis and the velocity vector rotate synchronously and eliminate sideslip as possible.

1. Lateral directional coordinated turn

In order to implement lateral-directional coordinated turn, the ailerons and rudder are always used. The sideslip angle is fed back to the rudder for the coordination control. See Figure 6.6 – 8.

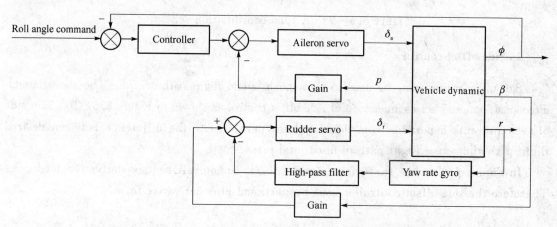

Figure 6.6 – 8 Block diagram of coordination control

CHAPTER 6 Flight Control of Missile

The roll rate p is fed back to the ailerons to modify the roll-subsidence mode. The yaw rate r is fed back to the rudder to modify the Dutch roll mode. If a high-pass filter $\tau s/(1+\tau s)$ is inserted, then the damper will vanish during steady-state conditions. The reason is that the high-pass filter produces an output only during the transient period. Thus, there is tradeoff between the enough damping improvement and the turn performance. The high-pass filter is sometimes known as the washout circuit. The ailerons deflection produces the roll rate. The roll rate causes the adverse yawing moment. The resulting sideslip occurs. It is naturally thought out that the sideslip angle is fed back to the rudder to eliminate sideslip. Sometimes, the lateral acceleration feedback is used instead of sideslip angle.

Figure 6.6 - 9 presents another scheme of turn-coordination. Introducing roll angle ϕ feedback to the aileron loop will increase the stability of spiral mode. Introducing yaw angular rate r feedback also improves the stability of Dutch roll mode. In addition, for coordinated turn, the roll angle ϕ is fed back to the rudder.

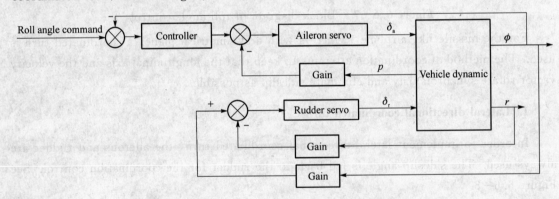

Figure 6.6 - 9   A turn-coordination scheme

2. Side offset control

Side offset control may be used to accomplish lateral flight path control. The above lateral-directional loop may act as inner loop of side offset control as shown in Figure 6.6 - 10. The side offset represents departure from desired flight path, i.e., the difference between desired flight path and actual flight path in horizontal plane.

In Figure 6.6 - 10, the offset rate feedback is equivalent to a derivative feedback. Therefore the side offset controller uses proportional plus derivative form.

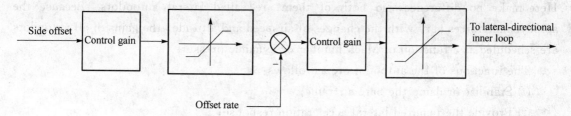

Figure 6.6 – 10 Block diagram of side offset control

6.7 Guidance System with Acceleration Control

When a missile is intercepting a moving target, an acceleration control loop commonly acts as inner loop of a guidance system. During interception, an on-board seeker senses the error of LOS. According to a guidance law, such as proportional navigation, a guidance command is generated. The acceleration closed loop implements guidance command to steer the missile flight at the target. A homing guidance system is illustrated in Figure 6.7 – 1.

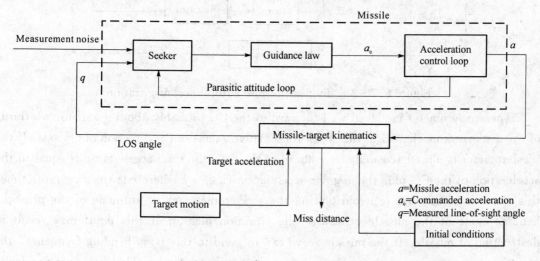

Figure 6.7 – 1 Block diagram of a homing guidance system

6.7.1 Acceleration Control

In general, there are three autopilots for a missile. They are roll autopilot, pitch autopilot and yaw autopilot respectively. Pitch autopilot is almost the same as yaw autopilot.

Here make no difference, so both of them are called lateral autopilots. Because the aerodynamic forces vary with the changes of airspeed and altitude, the gains of all autopilots are scheduled as a function of Mach number or dynamic pressure.

The functions of the autopilot are as follows:
① Stabilize or damp the bare airframe;
② Provide the required lateral acceleration response;
③ Reduce the sensitivity to disturbance within the missile's flight envelope.

Figure 6.7 - 2 shows a block diagram of generic autopilot, where an accelerometer provides the main feedback and a rate gyro is used as a damper. The typical acceleration control is commonly used for many high performance command and homing missiles. For the missile with two planes of symmetry, the pitching motion control is similar to the yawing motion control. It is enough to only consider one channel.

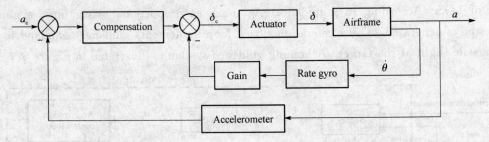

Figure 6.7 - 2 Typical missile autopilot configuration

The accelerometer is placed well forward of the CG probably about a half to two thirds of the distance from the CG to the nose. Its sensitive axis is in the direction of Oy axis. If the accelerometer is placed at a distance c ahead of the CG, the total acceleration is equal to the acceleration of the CG plus the angular acceleration (i. e. $\dot{r}$, where r is the yaw rate) times this distance c. One has to avoid placing the accelerometer at an antinode of the principal bending mode of the missile otherwise the vibration pick-up at this point may result in destruction of missile. If the missile servo can respond to this body bending frequency, the resulting fin movement will tend to reinforce this natural mode. The rate gyro is ideally placed at a node where the angular movement due to vibration is a minimum. The rate gyro's sensitive axis is along Oz, i. e. its output is proportional to yaw rate r.

Figure 6.7 - 3 shows the arrangement in transfer function form. The symbols in Figure 6.7 - 3 are defined as follows:
- a_{yd} = demanded lateral accelerations of the CG,
- a_y = achieved lateral accelerations of the CG,

6.7 Guidance System with Acceleration Control

- $-k_s =$ servo gain,
- $k_g =$ rate gyro gain,
- $c =$ distance of accelerometer sensitive axis forward of the CG.

The servo gain has to be negative since the transfer function a_y/δ_r has a negative gain.

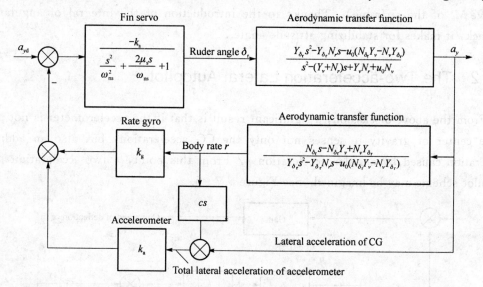

Figure 6.7-3 Lateral autopilot with one accelerometer and one rate gyro

For a static unstable missile, another control scheme may be used. See Figure 6.7-4.

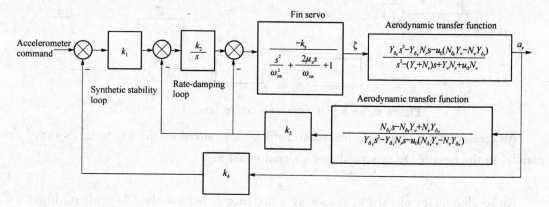

Figure 6.7-4 Pitch/yaw autopilot with synthetic stability loop

Referring to Figure 6.7-4, the introducing angular rate integral feedback (i.e. synthetic stability loop), which is equivalent to using incremental pitch angle feedback, may improve system performance. The inner-loop proportional feedback, combined with pure integral control, has the same characteristic equation as PI control but has eliminated the closed-loop PI zero.

The synthetic stability loop enables the autopilot to tolerate some instability (i. e. positive M_a of the airframe). Thanks to the introduction of the integral of angular rate feedback, it makes for stabilizing attitude angle.

6.7.2 The Two-acceleration Lateral Autopilot

From the above discussion, a significant result is that if the accelerometer is not placed at the center of gravity it senses not only the CG acceleration, but also an additional acceleration caused by angular acceleration $\dot{r}$. From this point, a two-acceleration lateral autopilot scheme may be imagined. See Figure 6.7-5.

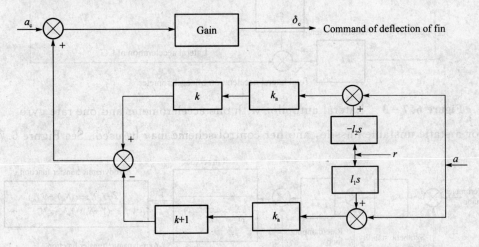

Figure 6.7-5 Two-acceleration lateral autopilot

An accelerometer of gain k_a placed at a distance l_1 ahead of the CG with its input axis parallel to the missile Oy axis produces a signal equal to

$$k_a(a_y + l_1 \dot{r}) \qquad (6.7-1)$$

An accelerometer of gain k_a placed at a distance l_2 behind the CG with its input axis parallel to the missile Oy axis produces a signal equal to

$$k_a(a_y - l_2 \dot{r}) \qquad (6.7-2)$$

If the forward accelerometer gain is increased to $(k+1)$ by an amplifier and the rear accelerometer gain is increased to k by another amplifier but this latter signal is fed back positively, then the resultant negative feedback is

$$(k+1)k_a(a_y+l_1\dot{r})-kk_a(a_y-l_2\dot{r})=k_a\{a_y+[l_1+(l_1+l_2)k]\dot{r}\} \quad (6.7-3)$$

This is equivalent to having one accelerometer much further forward of the CG.

By the way, the flexible use of two accelerometers brings about good design result. For example, if an inertial element $1/(Ts+1)$ is cascaded behind the rear accelerometer as shown in Figure 6.7-6, then the phase advance compensation can be obtained.

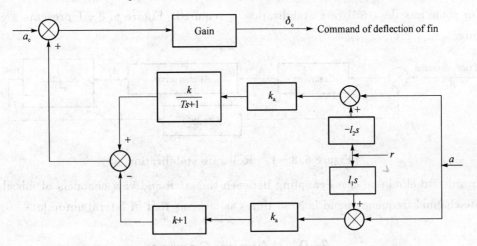

Figure 6.7-6 Two-acceleration lateral autopilot with a filter

The total negative feedback is

$$(k+1)k_a(a_y+l_1\dot{r})-\frac{k}{Ts+1}k_a(a_y-l_2\dot{r})=k_a\{a_y+[l_1+(l_1+l_2)k]\dot{r}\}=$$

$$k_a\frac{(k+1)Ts+1}{Ts+1}a_y+k_a\frac{(k+1)l_1Ts+(k+1)l_1+kl_2}{Ts+1}\dot{r}$$

This is equivalent to a $(k+1)$ to 1 phase advance network.

In the scheme, there are features as follows:

① If the chosen parameters are reasonable, one may obtain better dynamical qualities. But it should be emphasized that the transfer coefficient difference between the forward sensor and the rear sensor is one.

② It is convenient to accomplish lead compensation.

③ Line accelerometer has lower price, smaller volume, higher accuracy, and shorter preparing time than angular rate gyro.

In practice, it is specially noted that the CG change should locate between forward and rear accelerometers. If the CG position changes in front of the fore accelerometer, then the feedback becomes a positive feedback, and results in the system to be unstable; if the CG position changes behind the rear accelerometer, then the performances of the system become bad. In addition, since the missile is not absolute rigid body, the accelerometers are easy to sense elastic deformation, and result in the advent of adverse effects.

6.8 Roll Rate Stabilization

For some missiles, roll rate stabilization is required. Figure 6.8 – 1 presents a control structure.

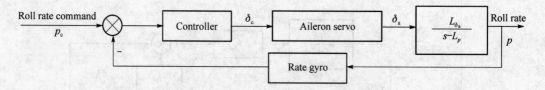

Figure 6.8 – 1 Roll rate stabilization

In order to eliminate cross-coupling between the pitch and yaw channels of missile, the roll rate channel frequency band is 3~5 times as large as that of lateral autopilot.

6.9 Missile Servos

According to control command, missile servo drives control surface (fin) to deflect or alter engine thrust vector direction so as to produce control force, which controls and stabilizes missile flight. The servo usually consists of a power amplifier, actuator, feedback element, operating mechanism, and fin as seen in Figure 6.9 – 1.

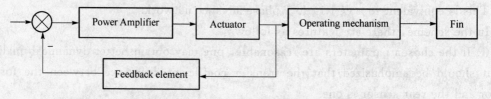

Figure 6.9 – 1 General principle block diagram of missile servo

In some small missiles, an open-loop structure is used without feedback component so as to make cost, and mass of this kind of structure be more reasonable.

In light of supplying energy sources, there are pneumatic, hydraulic, electric, and electromagnetic servos. They have advantages and drawbacks respectively.

6.9 Missile Servos

Basic requirements of servos are as follows.

1. Enough output moment

An actuator output moment M ought to satisfy

$$M \geqslant M_H + M_F + M_I \qquad (6.9-1)$$

where

M_H = hinge moment,
M_F = friction moment of fin and driving mechanism,
M_I = moment overcoming inertial moment of fin and driving mechanism.

2. Enough deflection angle and rate

This point is easily satisfied. In general, for tactical missile fin deflection angle is $15° \sim 20°$ and deflection rate is about $150(°)/s \sim 200(°)/s$. The mechanical limit of deflection angle has to be set.

3. Satisfactory dynamic response

The transient time is short. The overshoot is small. The frequency band is $3 \sim 7$ times as much as that of the stabilizing loop of airframe.

4. Small volume and light gravity

Servo requirements summarily depend on tactical and technical indexes of the missile.

6.9.1 Pneumatic Servos

Pneumatic servos include cold gas servo and hot gas servo. Higher pressure (typically about 5×10^6 N/m^2) gas from storage tank is used for energy source for the cold gas servo. Hot gas caused by the solid powder combustion is used to drive actuator motion.

1. Cold gas servos

A typical gas servo is indicated in Figure 6.9-2. Torque motor, which is controlled by output current of amplifier, deflects efflux nozzle to adjust intake flows on two sides of the cylinder. Depending on pressure difference of the two chambers, the piston motion is driven in the cylinder to implement fin deflection. Potentiometers are fixed at one end of piston pole. Brush of potentiometer moving with piston arm produces feedback signal to the amplifier.

Advantages of cold gas servo are clean and reliable. It may be stored in a long term. Moreover, the electronics and valve design are simple, and manufacture costs are low.

However, the gain and natural frequency of cold gas servo decrease with an increase in load inertia. Pneumatic servos are usually used for small missiles with fairly short flight time.

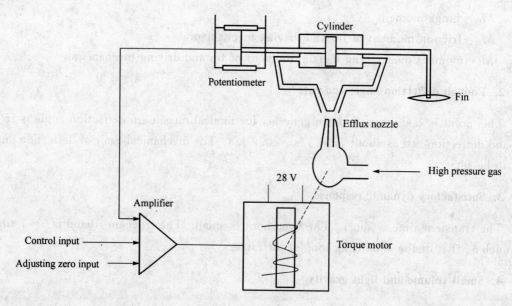

Figure 6.9 − 2   Principle block diagram of cold gas servo ("SA—2" surface-to-air missile)

2. Hot gas servos

A typical hot servo is shown in Figure 6.9 − 3.

When the missile launches, the ignitor ignites the powder. Hot gas, which is generated due to the power combustion is filtered, and assigned, acts on pistons through throttle apertures. Hot gas in assigning chamber passes through tilted aperture to drive worm wheel generator. The exhaust gas expels to air through outtake holes on the generator's shell.

Servo controller produces electric currents I_1, I_2 signal to electromagnetic coils, which generate pulling forces to attract bafflers. If $I_1 = I_2$, the two clearances between bafflers and spray nozzles are the same. As a result, pressures acting on two pistons are equal, and the fin does not deflect. If $I_1 \neq I_2$, the two clearances between the spray nozzles and bafflers are different. As a result a piston moves leftward (or rightward), while the other piston moves rightward (or leftward). Then the fin deflects until the driving moment balances hinge moment.

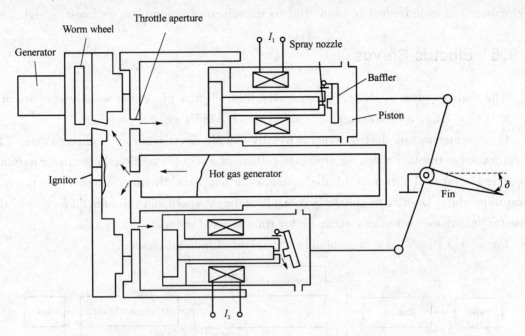

Figure 6.9 - 3 Hot gas servo (sidewinder missile)

6.9.2 Hydraulic Servos

Figure 6.9 - 4 shows a principle block diagram of hydraulic servo.

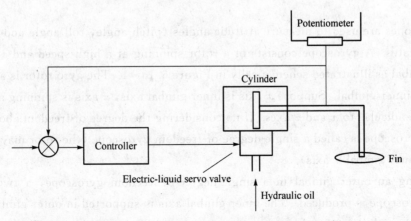

Figure 6.9 - 4 Principle block diagram of hydraulic servo

Hydraulic servo consists of a controller, feedback potentiometer, electric-liquid servo valve, cylinder, hydraulic source, and rocking arm. Hydraulic servo has the advantages of high-power and wide frequency band. But its manufacture is complex, and cost is high.

6.9.3 Electric Servos

The energy source of electric servo comes from electric power. It usually consists of a motor (DC or AC), a tachometer, a position sensor, and a gear.

Control manners of electric actuator usually include direct mode and indirect mode. The direct mode is, through changing armature voltage of a motor, to directly control rotational speed and rotational direction of the actuator output axis. Indirect manner is, through controlling clutch on-off, to indirectly control rotational speed and rotational direction of the actuator output axis when an electric motor runs at fixed speed.

Figure 6.9 – 5 shows a principle block diagram of indirect manner.

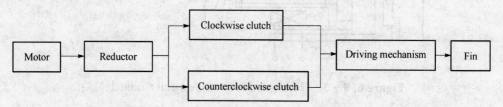

Figure 6.9 – 5 Principle block diagram of indirect manner

6.10 Gyroscopes

Gyroscopes are used to measure attitude angles (pitch angle, roll angle and yaw angle) or angular rates. A gyroscope consists of a rotor spinning at a high speed and a framework called a gimbal as illustrated schematically in Figure 6.10 – 1. The gyro rotor is supported in bearings in inner gimbal. Suppose x axis is inner gimbal axis, z axis is spinning axis, then y axis is perpendicular to x and z axes. If no considering the degree of freedom about spinning axis, the gyroscope is called a single-degree-of-freedom gyroscope (the rotor may rotate only about the inner gimbal x axis).

If adding an outer gimbal in a single-degree-of-freedom gyroscope, a two-degree-of-freedom gyroscope is produced. The inner gimbal axis is supported in outer gimbal, and the outer gimbal axis is supported in a base stand as seen in Figure 6.10 – 2. The gyroscope has

6.10 Gyroscopes

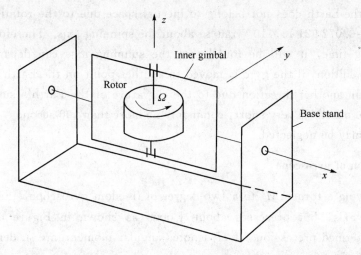

Figure 6.10 – 1 Single-degree-of-freedom gyroscope schematic diagram

two-degree-of-freedom, i.e., rotating about inner axis gimbal x and outer gimbal axis y.

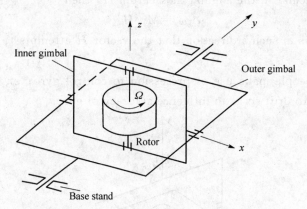

Figure 6.10 – 2 Two-degree-of-freedom gyroscope schematic diagram

Basic characteristics of gyroscopes are as follows.

1. Stability of gyroscope spinning axis orientation

When the gyro rotor spins at high speed Ω (generally about 21 500 r/min $\sim$ 24 000 r/min), if no external torque, which is not aligned with Ω, actuates the rotor, then the spinning axis pointing maintains stability with respect to inertial space no matter how the base stand rotates. Actually, there exist disturbance torques, and the torques result in drift of gyroscope. The drift causes measuring error.

However, the Earth does not belong to inertial space due to the rotation of the Earth (Earth rate $\omega_E = 0.072\,921\,15 \times 10^{-3}$ rad/s) about the spinning axis. Therefore, if observing the gyro in long time, it will be found that the spinning axis can depart from original orientation. In addition, if the gyro is moved to another point on the Earth's surface it will appear to point in another direction due to the curvature of the Earth's surface. Since the operating time of missile is very short, commonly no more than 100 seconds, the effect of the Earth spinning may be neglected.

2. Precession of gyroscope

When applying a torque M_x to a two-degree-of-freedom gyroscope, the rotor does not rotate about x axis, but precesses about y axis as shown in Figure 6.10 – 3. The phenomenon is termed precession. If H denotes angular momentum, ω_y denotes precession rate, then their relation is given by

$$\omega_y \times H = M_x \qquad (6.10-1)$$

If ω_y is perpendicular to the angular momentum H, then

$$\omega_y = M_x / H \qquad (6.10-2)$$

The precession is in such a direction that the vector H attempts to align itself with the applied torque vector.

On one hand, people make use of precession to control gyro; on the other hand, the precession causes gyro drift so as to influence accuracy of gyro.

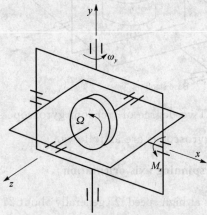

Figure 6.10 – 3 Precession of gyroscope schematic diagram

3. Gyroscope moment

The Newton's third law points that action force and reaction force are equal in magnitude, opposite in direction, and act on two different objects respectively. Thus, the gyro precession inevitably creates a reaction moment to oppose any applied moment. The reaction moment is called the gyroscope moment. If M_g denotes the gyroscope moment, M_{app} denotes the applied moment, ω denotes the angular velocity of precession, and H denotes angular momentum, then

$$M_g = -M_{app} = H \times \omega \qquad (6.10-3)$$

6.11　Free or Position Gyroscopes

Figure 6.11-1 shows a position gyroscope, which has two degrees of freedom, can measure two attitude angles such as roll angle and pitch angle. The base of potentiometer of roll is fixed on outer gimbal. Its sliding arm is connected with the axis of inner gimbal. The base of potentiometer of pitch is fixed on the body of missile. Its sliding arm is connected with the axis of outer gimbal. If the missile rolls angle ϕ about the x axis, then the outer gimbal and the base of potentiometer of roll rotate ϕ about the x axis with the missile. The inner gimbal and the sliding arm of potentiometer of roll remain unchanged in space position. As a result, the potentiometer of roll gives a voltage which is proportional to the roll angle ϕ. In the same manner, the potentiometer of pitch gives a voltage which is proportional to the pitch angle θ.

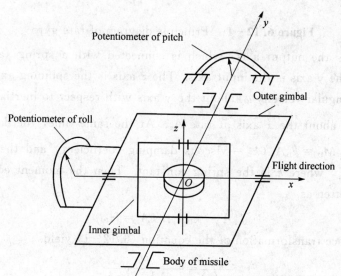

Figure 6.11-1　Position gyro schematic diagram

CHAPTER 6 Flight Control of Missile

Obviously, if the pitch angle, roll angle, and yaw angle are required to be measured at the same time, it is necessary to set two gyros, whose spinning axes are mutually vertical.

6.12 Rate Gyroscopes

Figure 6.12-1 shows the configuration of a typical rate gyro.

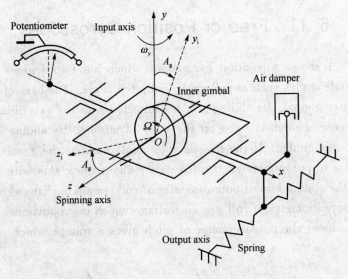

Figure 6.12-1 Principle diagram of rate gyro

The x axis is the output axis, which is connected with a spring, an air damper, a potentiometer. The y axis is the input axis. The z axis is the spinning axis. If the base of the gyro has an angular velocity ω_y about the y axis with respect to inertial space, then the gyro will precess about the x axis at rate $\dot{A}_g$. At the same time, the damper produces a damping moment $M_D = k_D \dot{A}_g$ (k_D = viscous damping coefficient) and the spring produces moment $M_S = k A_g$, where k is the spring constant. Then the moment equation about the output axis is written as

$$J_x \ddot{A}_g + k_D \dot{A}_g + k A_g = H \omega_y \qquad (6.12-1)$$

Taking Laplace transformation of the equation (6.12-1) yields

$$\frac{A_g(s)}{\omega_y(s)} = \frac{H}{J_x s^2 + k_D s + k} \qquad (6.12-2)$$

In steady state, i.e., $s = 0$, then the steady angle A_g^* is

$$A_g^* = \frac{H}{k}\omega_y \qquad (6.12-3)$$

In order to give the readers some conception of the magnitudes of the various parameters of a typical rate gyro, an example is given. For a rate gyro, there are the following parameters.

$H=10^4$ g·cm²/s, $k_D=5\times 10^3$ g·cm²/s, $k=3.03\times 10^5$ g·cm²/s², $J_x=34$ g·cm²

Then the natural frequency, damping ratio and gain are computed as follows:

$$\omega_n = \sqrt{k/J_x} = \sqrt{3.03\times 10^5/34} \text{ rad/s} = 94.4 \text{ rad/s}$$

$$\xi = k_D/(2\sqrt{kJ_x}) = 5\times 10^3/(2\times\sqrt{3.03\times 10^5\times 34}) = 0.779$$

$$H/k = 10^4/(3.03\times 10^5) \text{ s} = 0.033 \text{ s}$$

In practice, there is friction in motion of gyro. It is regarded that the magnitude of the friction moment M_F is unchanged, but its direction is contrary to $\dot{A}_g$, i.e.

$$M_F = -M_{Fx}\text{sgn }\dot{A}_g \qquad (6.12-4)$$

Only if $H\omega_y \geqslant M_{Fx}$, the inner gimbal can rotate. Therefore, sensible minimum angular speed $\omega_{y,\min}$ is

$$\omega_{y,\min} = \frac{M_{Fx}}{H} \qquad (6.12-5)$$

In order to measure small angular velocity, it is necessary to decrease friction moment, or increase angular moment H. For example, if $1(°)/h$ angular velocity is expected to be measured, then one may evaluate permitted value of friction moment. Supposing angular moment of gyro $H=4\,000$ g·cm²/s, then

$$M_{Fx} = 4\,000\times 1\times \frac{1}{60\times 60\times 57.3} \text{ g·cm} = 0.019\,4 \text{ g·cm}$$

For a common bearing, its friction moment is more than 0.02 g·cm. The floated gyroscope reaches the requirement.

6.13 Accelerometers

The accelerometer is used to measure acceleration of missile motion. Figure 6.13-1 illustrates an accelerometer, which consists of a sensing mass block, damper, spring, and potentiometer.

If the displacement of a vehicle with respect to inertial space is x_i, corresponding acceleration $a_x = \dfrac{d^2 x_i}{dt^2}$, then the displacement of the accelerometer's case of is also x_i, and the

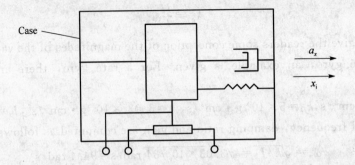

Figure 6.13-1 Accelerometer principle schematic diagram

acceleration of the case is a_x because the case is fixed on the vehicle. Assuming the displacement of sensing mass block with respect to inertial space is x_m, then the displacement of the sliding arm with respect to the case is $x = x_m - x_i$.

From Figure 6.13-1 it may be seen that the differential equation of motion for the mass m is

$$m\frac{d^2 x_m}{dt^2} + k_D \frac{dx}{dt} + kx = 0 \qquad (6.13-1)$$

where

m = mass of the mass block,
k_D = damping coefficient,
k = stiffness coefficient of the spring.

Substituting $x_m = x_i + x$ into the equation (6.13-1) yields

$$-m\frac{d^2 x_i}{dt^2} = m\frac{d^2 x}{dt^2} + k_D \frac{dx}{dt} + kx \qquad (6.13-2)$$

i. e.

$$-ma_x = m\frac{d^2 x}{dt^2} + k_D \frac{dx}{dt} + kx \qquad (6.13-3)$$

Taking Laplace transformation of the equation (6.13-3) yields

$$\frac{x(s)}{a_x(s)} = -\frac{m}{ms^2 + k_D s + k} = -\frac{1}{s^2 + \frac{k_D}{m}s + \frac{k}{m}} \qquad (6.13-4)$$

In steady state, $\frac{d^2 x}{dt^2} = \frac{dx}{dt} = 0$, the equation (6.13-3) becomes

$$a_x = -\frac{k}{m}x \qquad (6.13-5)$$

The displacement of the mass block is proportional to the acceleration of the vehicle.

The potentiometer produces a voltage signal proportional to the acceleration of the vehicle.

Figure 6.13 - 2 shows a liquid-floated pendulous accelerometer. The case of the accelerometer is filled with liquid, where the cylindrical floater floats. In the floater there is a bias mass m, which has a distance l from x axis.

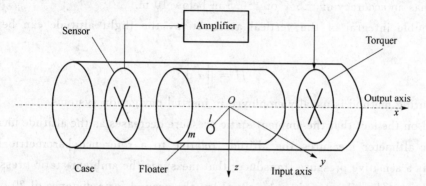

Figure 6.13 - 2   Liquid-floated pendulous acceleration principle schematic diagram

If the case has acceleration a along y axis, the mass block m produces an inertial moment about x axis due to inertial action as follows

$$M_i = lma \tag{6.13-6}$$

Under the action of the inertial moment, the rotor of the signal sensor rotates an angle θ. The sensor gives a voltage $u = k_\theta \theta$ proportional to θ. The amplifier produces a current $i = k_u u$ proportional to u. It is fed to the torquer. The torquer creates a moment $M_t = k_i i$ proportional to i. In steady state, the moment balances the inertial moment, i.e.

$$M_t = k_i i = lma \tag{6.13-7}$$

Thus,

$$i = \frac{lma}{k_i} \tag{6.13-8}$$

Here, the input current of the torquer is proportional to input acceleration. Through sampling resistance, a voltage signal proportional to the acceleration can be obtained.

6.14   Altimeters

Flight altitude is a vertical distance from the vehicle to the reference plane. For a vehicle, different reference planes correspond to different altitude. An altimeter is a device measuring the altitude of a vehicle. There are three methods measuring altitude in common.

The electromagnetic wave is used to measure the flight altitude. Through measuring return time from a vehicle to ground, the radio altimeter indirectly obtains altitude signal, i. e.

$$H = c \frac{\Delta t}{2} \qquad (6.14-1)$$

where, c is the transmitting speed of electromagnetic wave. A typical commercial FM/CW altimeter has an accuracy of $\pm 5\%$ or ± 0.5 m below 10 m.

By double integral of the vertical acceleration, the flight altitude can be obtained. That is

$$H = \iint a_z \mathrm{d}t^2 \qquad (6.14-2)$$

The measuring method is usually employed in inertial navigation systems.

Based on the fact that the ambient static pressure decreases as the altitude increases, the barometric altimeter measures the altitude relative to a reference barometric level. The altimeter is a sensitive pressure transducer that measures the ambient static pressure. If the missile is required to fly at a given height above the ground for a distance of 20 or 30 km, a simple barometric capsule or even a piezo electric pressure transducer should be accurate enough to indicate height. Below 100 m, the error of the barometric altimeter is large.

References

[1] Garnell P. Guided weapon Control Systems. Oxford:Pergamon Press,1980.
[2] George M Siouris. Missile Guidance and Control Systems. New york: Springer,2003.
[3] 俞超志. 导弹概论. 北京:北京工业学院出版社, 1986.
[4] 钱杏芳, 林瑞雄, 赵亚男. 导弹飞行力学. 北京:北京理工大学出版社,2000.
[5] 祝小平. 无人机设计手册. 北京: 国防工业出版社,2007.
[6] 娄寿春. 导弹制导技术. 北京:中国宇航出版社, 1989.

CHAPTER 7
Guidance Laws

When a missile approaches a target, a motion law which should be obeyed by the missile is defined as guidance law. Supposing that the target, the missile, and the guidance station are particles, the flight path, which pre-arranged guidance law determine, is called a kinematical trajectory. The kinematical trajectory rests with guidance law and target motion law. In general, even if a target does not maneuver, guidance trajectory may be curved. Virtually, a missile is not a particle, but has size and shape. The missile motion can not instantaneously finish due to inertia. Hence, the missile moves along dynamical trajectory. In addition, during flight, there are disturbances. In fact, there are not only disturbances, but also instrumental errors, so the missile moves along practical trajectory.

Guidance laws determined by a missile and a target positions in space include pursuit method, proportional navigation, and constant bearing guidance. These guidance methods are commonly used in homing guidance systems. Guidance laws determined by a missile and target and guidance station's positions in space include three-point method and lead angle method. The two guidance methods are used in remote control guidance systems. When designing a guidance law, people usually select smaller curvature trajectory by reason that: ① the more straight the shape of the trajectory, the shorter the journey of the flight will be, and as a result, the flight time will be shortened; ② the more straight the shape of the trajectory, the lower the maneuverability requirement for the missile will be, and as a result the requirement for structural strength decreases.

In order to discuss guidance laws, the motion of a target is firstly discussed.

7.1 Motion of a Target

When use a radar to detect a target T, the polar coordinates of the target $(R_T, \varepsilon_T, \beta_T)$

CHAPTER 7 Guidance Laws

are obtained. See Figure 7.1-1. Here, suppose point O_E is chosen in guidance station location. The $O_E x_E$ axis points to north direction. The $O_E z_E$ axis points vertically downward. The $O_E y_E$ axis forms a right handed orthogonal system with the other two axes. R_T is a vector radius of the target. $|R_T|$ is the slant range of the target. The elevation angle ε_T is an angle between the vector radius of the target R_T and the $O_E y_E y_E$ plane. The azimuth β_T is an angle between the projection of R_T on the $O_E x_E y_E$ plane and the $O_E x_E$ axis. If the projection is rightward reference to the $O_E x_E$ axis, the angle is positive.

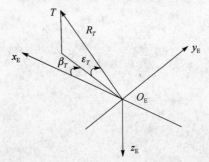

Figure 7.1-1 The polar coordinates of the target $(R_T, \varepsilon_T, \beta_T)$

It is easy to obtain orthogonal coordinates with the target as follows:

$$\left.\begin{array}{l} x_T = R_T \cos \varepsilon_T \cos \beta_T \\ y_T = R_T \cos \varepsilon_T \sin \beta_T \\ z_T = -R_T \sin \varepsilon_T \end{array}\right\} \quad (7.1-1)$$

1. The radar's detecting parameters for a target of uniform rectilinear motion

Assume the target to fly at constant velocity V_T to a guidance station as seen in Figure 7.1-2. $TA = H_T$; $AB = D_T$; $O_E T = R_T$. Suppose flight altitude H_T and crossing distance D_T to be positive constants. Let us discuss motion parameters.

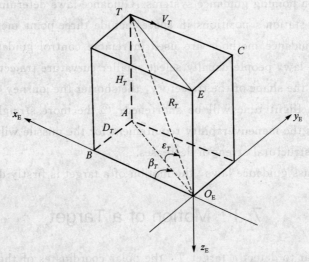

Figure 7.1-2 Geometrical relation of target motion

(1) Closing velocity $\dot{R}_T$ and acceleration $\ddot{R}_T$

From Figure 7.1-2,

$$\dot{R}_T = -V_T \cos\beta_T \cos\varepsilon_T = -\frac{\sqrt{R_T^2 - D_T^2 - H_T^2}}{R_T} V_T \qquad (7.1-2)$$

$$\ddot{R}_T = V_T(\dot{\beta}_T \sin\beta_T \cos\varepsilon_T + \dot{\varepsilon}_T \cos\beta_T \sin\varepsilon_T) = \frac{D_T^2 + H_T^2}{R_T^3} V_T^2 \qquad (7.1-3)$$

(2) Azimuth angular velocity $\dot{\beta}_T$ and acceleration $\ddot{R}_T$

$$\dot{\beta}_T = \frac{V_T \sin\beta_T}{O_E A} = \frac{V_T D_T / O_E A}{O_E A} = \frac{V_T D_T}{R_T^2 - H_T^2} \qquad (7.1-4)$$

$$\ddot{R}_T = \frac{V_T^2}{R_T^2 \cos^2\varepsilon_T} \sin 2\beta_T \qquad (7.1-5)$$

An observation of the equation (7.1-4) shows that when the target is closing to the guidance station, the R_T decreases and the $\dot{\beta}_T$ will increase. When the target reaches the point C, $\dot{\beta}_T$ has maximum value. On the other hand, if the target evades from the guidance, the azimuth angular velocity $\dot{\beta}_T$ will decrease. At point C, $\ddot{R}_T = 0$.

(3) Elevation angular velocity $\dot{\varepsilon}_T$ and acceleration $\ddot{\varepsilon}_T$

$$\dot{\varepsilon}_T = \frac{V_T \cos\beta_T \sin\varepsilon_T}{R_T} \qquad (7.1-6)$$

$$\ddot{\varepsilon}_T = \frac{V_T^2}{R_T^2}(2\cos^2\beta_T \cos^2\varepsilon_T - \sin^2\beta_T)\tan\varepsilon_T = $$

$$\frac{V_T^2}{R_T^2}[\cos^2\beta_T(1 + 2\cos\varepsilon_T) - 1]\tan\varepsilon_T \qquad (7.1-7)$$

If $D_T \neq 0$, then at point $C, \beta_T = 90°$, i.e. $\cos\beta_T = 0$, $\dot{\varepsilon}_T = 0$.

If $D_T = 0$, i.e. $\beta_T = 0$, then at point E, $\dot{\varepsilon}_T$ has maximum value $\frac{V_T}{R_T}$, here $\varepsilon_T = 90°$.

Through above analyses, one knows that even though a target moves at constant speed and constant altitude, the target's parameters $(R_T, \varepsilon_T, \beta_T)$, which are measured by the radar, are time-varying and nonlinear.

2. Target maneuver

In order to avoid being attacked, a target often maneuvers. Here, the target can produce acceleration.

$$\boldsymbol{a}_T = \frac{d\boldsymbol{V}_T}{dt} = \frac{d}{dt}(V_T \boldsymbol{V}°) = \frac{dV_T}{dt}\boldsymbol{V}° + \frac{d\boldsymbol{V}°}{dt}V_T = \boldsymbol{a}_{Tt} + \boldsymbol{a}_{Tn} \qquad (7.1-8)$$

where

$V°=$ unit vector of target velocity,

$a_{Tt} = \dfrac{dV_T}{dt} V° =$ tangent acceleration of target,

$a_{Tn} = \dfrac{dV°}{dt} V_T =$ normal acceleration of target.

The magnitude of flight velocity depends on the tangent acceleration, but the direction of flight velocity rests with the normal acceleration.

Thus there are two maneuvering forms: ① changing of speed magnitude maneuver and ② changing of direction maneuver. The load factor may be used to represent target maneuver. For example, if a target maneuvers in vertical plane, then the acceleration of the target is

$$a_z = n_z g \qquad (7.1-9)$$

In the same manner, a target may maneuver in horizontal plane. Nowadays the load factor of a fighter aircraft reaches 9. If a target adopts S-shaped maneuver in horizontal plane, then the acceleration is

$$a_y = n_y g \cos \omega t \qquad (7.1-10)$$

For example the load factor of anti-ship missile attains $10 \sim 15$, and the period of S-shaped maneuver is $3 \sim 6$ s.

If a target produces two accelerations in horizontal plane and in vertical plane, i.e.

$$\left. \begin{array}{l} a_z = n_z g \cos \omega t \\ a_y = n_y g \sin \omega t \end{array} \right\} \qquad (7.1-11)$$

then the target executes space lateral roll maneuver. When $n_y = n_z$, the target maneuvers with uniform lateral roll. In order to intercept high maneuver targets, missiles have to have higher maneuverability. For example, an anti-radiation missile flies at $Ma = 3$. Its load factor of space lateral roll reaches 20, and circular frequency ω is $2 \sim 3$ rad/s.

7.2 Remote Control Guidance Method

For remote control guidance, the guidance station detects motion parameters of missiles and targets. According to the missile's position relative to line of sight from the guidance station to the target, there are three-point method and lead angle method.

Suppose the target T and missile M fly in the same vertical plane as shown in Figure 7.2-1.

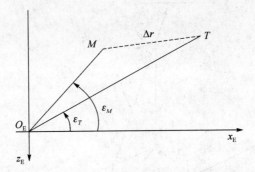

Figure 7.2-1 Geometrical relation of remote control guidance

For remote control guidance, the guidance equations are

$$\varepsilon_M = \varepsilon_T + \Delta r \cdot f(t) \qquad (7.2-1)$$
$$\beta_M = \beta_T + \Delta r \cdot f(t) \qquad (7.2-2)$$

where

ε_M = elevation angle of missile,
ε_T = elevation angle of target,
β_M = azimuth angle of missile,
β_T = azimuth angle of target,
$\Delta r = R_T - R_M$ = missile-target relative distance,
$f(t)$ = lead quantity function.

7.2.1 Three-point Method

Three-point method requires that a guidance station and missile and target sustain a straight line in guidance process. This implies $f(t) = 0$ in equations (7.2-1) and (7.2-2). Then

$$\varepsilon_M = \varepsilon_T \qquad (7.2-3)$$
$$\beta_M = \beta_T \qquad (7.2-4)$$

In order to discuss three-point method an observing coordinate system is defined as illustrated in Figure 7.2-2. Original point O is located at guidance station. Ox axis points at the target. Oz axis lies in plumb plane, and is perpendicular to Ox axis. Oy axis forms a right handed system with the other axes. Suppose the target and the missile lie at point T and point M in x axis respectively. Assume unit vectors in $Oxyz$ system are i, j, and k respectively.

CHAPTER 7 Guidance Laws

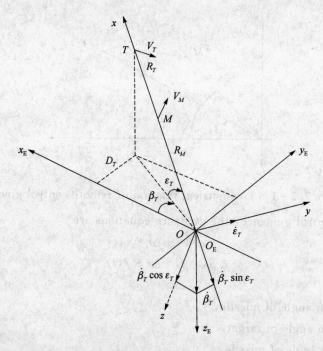

Figure 7.2-2 Geometrical relation of three-point method

If the elevation angle of the target and the azimuth angle of the target are denoted by ε_T, β_T, then the angular velocity vector $\boldsymbol{\omega}_T$ in $Oxyz$ system is

$$\boldsymbol{\omega}_T = -\dot{\beta}_T \sin \varepsilon_T \boldsymbol{i} + \dot{\varepsilon}_T \boldsymbol{j} + \dot{\beta}_T \cos \varepsilon_T \boldsymbol{k} \tag{7.2-5}$$

The velocity vector of the missile is

$$\boldsymbol{V}_M = \frac{d\boldsymbol{R}}{dt} = \frac{d(R_M \boldsymbol{i})}{dt} = \dot{R}_M \boldsymbol{i} + R_M \dot{\boldsymbol{i}} \tag{7.2-6}$$

Because

$$\dot{\boldsymbol{i}} = \begin{vmatrix} \boldsymbol{i} & \boldsymbol{j} & \boldsymbol{k} \\ -\dot{\beta}_T \sin \varepsilon_T & \dot{\varepsilon}_T & \dot{\beta}_T \cos \varepsilon_T \\ 1 & 0 & 0 \end{vmatrix} = \dot{\beta}_T \cos \varepsilon_T \boldsymbol{j} - \dot{\varepsilon}_T \boldsymbol{k} \tag{7.2-7}$$

the acceleration of the missile is

$$\boldsymbol{a}_M = \dot{\boldsymbol{V}}_M = \frac{d}{dt}(\dot{R}_M \boldsymbol{i} + R_M \dot{\boldsymbol{i}}) =$$

$$\ddot{R}_M \boldsymbol{i} + \dot{R}_M \dot{\boldsymbol{i}} + \dot{R}_M \dot{\boldsymbol{i}} + R_M \ddot{\boldsymbol{i}} = \ddot{R}_M \boldsymbol{i} + 2\dot{R}_M \dot{\boldsymbol{i}} + R_M \ddot{\boldsymbol{i}} =$$

$$\ddot{R}_M \boldsymbol{i} + 2\dot{R}_M (\dot{\beta}_T \cos \varepsilon_T \boldsymbol{j} - \dot{\varepsilon}_T \boldsymbol{k}) + R_M \frac{d}{dt}(\dot{\beta}_T \cos \varepsilon_T \boldsymbol{j} - \dot{\varepsilon}_T \boldsymbol{k}) =$$

$$\ddot{R}_M \boldsymbol{i} + 2\dot{R}_M(\dot{\beta}_T\cos\varepsilon_T\boldsymbol{j} - \dot{\varepsilon}_T\boldsymbol{k}) + R_M[(\ddot{\beta}_T\cos\varepsilon_T - \dot{\beta}_T\sin\varepsilon_T\dot{\varepsilon}_T)\boldsymbol{j} +$$
$$\dot{\beta}_T\cos\varepsilon_T\dot{\boldsymbol{j}} - \ddot{\varepsilon}_T\boldsymbol{k} - \dot{\varepsilon}_T\dot{\boldsymbol{k}}] \quad (7.2-8)$$

where

$$\dot{\boldsymbol{j}} = \begin{vmatrix} \boldsymbol{i} & \boldsymbol{j} & \boldsymbol{k} \\ -\dot{\beta}_T\sin\varepsilon_T & \dot{\varepsilon}_T & \dot{\beta}_T\cos\varepsilon_T \\ 0 & 1 & 0 \end{vmatrix} = -\dot{\beta}_T\cos\varepsilon_T\,\boldsymbol{i} - \dot{\beta}_T\sin\varepsilon_T\boldsymbol{k} \quad (7.2-9)$$

$$\dot{\boldsymbol{k}} = \begin{vmatrix} \boldsymbol{i} & \boldsymbol{j} & \boldsymbol{k} \\ -\dot{\beta}_T\sin\varepsilon_T & \dot{\varepsilon}_T & \dot{\beta}_T\cos\varepsilon_T \\ 0 & 0 & 1 \end{vmatrix} = \dot{\varepsilon}_T\boldsymbol{i} + \dot{\beta}_T\sin\varepsilon_T\boldsymbol{j} \quad (7.2-10)$$

Substituting equations (7.2-9) and (7.2-10) into (7.2-8) yields

$$\boldsymbol{a}_M = (\ddot{R}_M - R_M\dot{\varepsilon}_T^2 - R_M\dot{\beta}_T^2\cos^2\varepsilon_T)\boldsymbol{i} +$$
$$(2\dot{R}_M\dot{\beta}_T\cos\varepsilon_T + R_M\ddot{\beta}_T\cos\varepsilon_T - 2R_M\dot{\beta}_T\dot{\varepsilon}_T\sin\varepsilon_T)\boldsymbol{j} +$$
$$(-2\dot{R}_M\dot{\varepsilon}_T - R_M\ddot{\varepsilon}_T - R_M\dot{\beta}_T^2\cos\varepsilon_T\sin\varepsilon_T)\boldsymbol{k} \quad (7.2-11)$$

If the tangent acceleration of the missile is denoted by $\boldsymbol{a}_t$, and the unit vector of missile velocity V_M is denoted by $\boldsymbol{V}_M^\circ$, then

$$\boldsymbol{a}_t = \dot{V}_M\boldsymbol{V}_M^\circ = \dot{V}_M\frac{\boldsymbol{V}_M}{V_M} = \frac{\dot{V}_M}{V_M}\dot{\boldsymbol{R}}_M = \frac{\dot{V}_M}{V_M}(\dot{R}_M\boldsymbol{i} + R_M\dot{\boldsymbol{i}}) =$$
$$\frac{\dot{V}_M}{V_M}(\dot{R}_M\boldsymbol{i} + R_M\dot{\beta}_T\cos\varepsilon_T\boldsymbol{j} - R_M\dot{\varepsilon}_T\boldsymbol{k}) \quad (7.2-12)$$

If the lateral acceleration of the missile is denoted by $\boldsymbol{a}_n$, then

$$\boldsymbol{a}_n = \boldsymbol{a}_M - \boldsymbol{a}_t \quad (7.2-13)$$

Substituting the equations (7.2-11) and (7.2-12) into (7.2-13) yields

$$\boldsymbol{a}_n = \left(-\frac{\dot{V}_M}{V_M}\dot{R}_M + \ddot{R}_M - R_M\dot{\varepsilon}_T^2 - R_M\dot{\beta}_T^2\cos^2\varepsilon_T\right)\boldsymbol{i} +$$
$$\left[\left(2\dot{R}_M - \frac{\dot{V}_M}{V_M}R_M\right)\dot{\beta}_T\cos\varepsilon_T + R_M(\ddot{\beta}_T\cos\varepsilon_T - 2\dot{\beta}_T\dot{\varepsilon}_T\sin\varepsilon_T)\right]\boldsymbol{j} +$$
$$\left[\left(-2\dot{R}_M + \frac{\dot{V}_M}{V_M}R_M\right)\dot{\varepsilon}_T - R_M(\ddot{\varepsilon}_T + \dot{\beta}_T^2\cos\varepsilon_T\sin\varepsilon_T)\right]\boldsymbol{k} \quad (7.2-14)$$

Components a_{ny}, a_{nz} of $\boldsymbol{a}_n$ in observing coordinate system are respectively

$$a_{ny} = \left(2\dot{R}_M - \frac{\dot{V}_M}{V_M}R_M\right)\dot{\beta}_T\cos\varepsilon_T + R_M(\ddot{\beta}_T\cos\varepsilon_T - 2\dot{\beta}_T\dot{\varepsilon}_T\sin\varepsilon_T) \quad (7.2-15)$$

CHAPTER 7 Guidance Laws

$$a_{nz} = \left(-2\dot{R}_M + \frac{\dot{V}_M}{V_M}R_M\right)\dot{\varepsilon}_T - R_M(\ddot{\varepsilon}_T + \dot{\beta}_T^2 \cos\varepsilon_T \sin\varepsilon_T) \qquad (7.2-16)$$

If the velocity of the target is not fast, then the $\ddot{\beta}_T$, $\dot{\beta}_T\dot{\varepsilon}_T$, $\ddot{\varepsilon}_T$, and $\dot{\beta}_T^2$ are small. Thus the equations (7.2-15) and (7.2-16) are approximately

$$a_{ny} = \left(2\dot{R}_M - \frac{\dot{V}_M}{V_M}R_M\right)\dot{\beta}_T \cos\varepsilon_T \qquad (7.2-17)$$

$$a_{nz} = \left(-2\dot{R}_M + \frac{\dot{V}_M}{V_M}R_M\right)\dot{\varepsilon}_T \qquad (7.2-18)$$

From equation (7.2-17), it can be seen that the a_{ny} is proportional to $\dot{\beta}_T \cos\varepsilon_T$. From equation (7.2-18), the a_{nz} is proportional to $\dot{\varepsilon}_T$. In head-on attack, the $\dot{\beta}_T$ gradually increases, required acceleration a_{ny} also increases and the trajectory becomes more and more curved. Inversely, in stern-chase attack, the $\dot{\beta}_T$ decreases gradually, the required acceleration a_{ny} also decreases and the trajectory becomes less and less curved.

Example 7.2-1: Three-point guidance of anti-tank missile. Figure 7.2-3 shows that a tank moves at constant velocity $V_T = 12$ m/s along a non-maneuvering straight-line course. An anti-tank missile intercepts the target with three-point method. The missile speed V_M is constant, $V_M = 120$ m/s. Initial intercepting conditions are

- initial slant range of the target from the guidance station O, $R_{T0} = 3\,000$ m,
- initial slant range of the missile from the guidance station O, $R_{M0} = 50$ m,
- angle of the initial line of sight $q_{M0} = q_{T0} = 70°$.

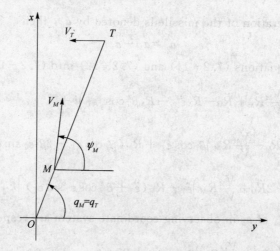

Figure 7.2-3 Three-point method guidance of an anti-tank missile

The motion equations are written as

$$\left.\begin{aligned}\frac{dR_M}{dt} &= V_M \cos(q_M - \psi_M) \\ \frac{dq_M}{dt} &= -\frac{V_M}{R_M} \sin(q_M - \psi_M) \\ \frac{dR_T}{dt} &= -V_T \cos q_T \\ \frac{dq_T}{dt} &= \frac{V_T}{R_T} \sin q_T \\ q_M &= q_T \end{aligned}\right\} \quad (7.2-19)$$

For integral convenience, the equations (7.2 – 19) are rewritten as

$$\left.\begin{aligned}\psi_M &= q_M + \arcsin\left(\frac{V_T R_M}{V_M R_T} \sin q_M\right) \\ \frac{dR_M}{dt} &= V_M \cos(q_M - \psi_M) \\ \frac{dR_T}{dt} &= -V_T \cos q_T \\ \frac{dq_M}{dt} &= -\frac{V_M}{R_M} \sin(q_M - \psi_M) \end{aligned}\right\} \quad (7.2-20)$$

These are nonlinear differential equations. In general, it is difficult to solve the equations. Fortunately, MATLAB has a built-in function that performs Runge-Kutta integral. It is very convenient to use the function. The Runge-Kutta integrator is called ode45. When applying ode45, it requires a MATLAB M file containing the nonlinear differential equations. MATLAB program 7.2 – 1 gives the M file.

In order to plot missile and target courses, it is necessary to give the following equations.

$$\left.\begin{aligned} x_M &= R_M \cos q_M \\ y_M &= R_M \sin q_M \end{aligned}\right\} \quad (7.2-21)$$

$$\left.\begin{aligned} x_T &= R_T \cos q_T \\ y_T &= R_T \sin q_T \end{aligned}\right\} \quad (7.2-22)$$

CHAPTER 7 Guidance Laws

MATLAB program 7.2 – 1

% Three-point method guidance for an anti-tank missile

function xdot=tpm(t,x)
% RM=x(1); ! Missile slant range
% QM=x(2); ! Missile line of sight angle in radian
% RT=x(3); ! Target slant range

VM=120;   % Missile speed
VT=12;   % Target speed

phi=x(2)+asin((VT/VM)*x(1)*sin(x(2))/x(3));
xdot(1)=VM*cos(x(2)−phi);
xdot(2)=−VM*sin(x(2)−phi)/x(1);
xdot(3)=−VT*cos(x(2));
xdot=xdot';

Entering MATALB program 7.2 – 2 in the command window of MATALB produces a missile's course as shown in Figure 7.2 – 4.

MATLAB program 7.2 – 2

```
>> [t,x]=ode45('tpm',[0 24],[50 70/57.29578,3000]);
>> [m,n]=size(x(:,:,:));
>> for i=1:m
        xm(i)=x(i,1)*sin(x(i,2));
        ym(i)=x(i,1)*cos(x(i,2));
        xt(i)=x(i,3)*sin(x(i,2));
        yt(i)=x(i,3)*cos(x(i,2));
    end
>> plot(ym,xm,'black',yt,xt,'black');
>> xlabel('y (m)');
>> ylabel('x (m)');
```

Now estimate the impact time. Since the impact point lies on the course of target, corresponding coordinate of the missile in x axis at impact point is

$$x_{Mf}=R_{T0}\sin q_{T0}=3\ 000\ \text{m}\times 0.939\ 7=2\ 819.1\ \text{m}$$

The interpolation is used to evaluate the impact time t_f. Entering MATLAB program 7.2-3 produces t_f.

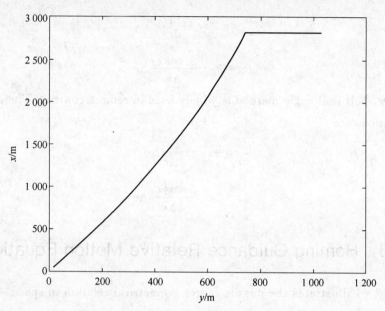

Figure 7.2-4 Missile and target courses of three-point guidance

```
MATLAB program 7.2-3
>> x0=xm';
>> y0=ym';
>> ymf=interp1(x0,y0,2819.1,'cubic');
>> tf=(3000*cos(70/57.29578)-ymf)/12
tf =
   23.90652565496008
```

Thus the total elapsed time of the missile from launching to impact the target is about $t_f = 23.9$ s.

7.2.2 Lead Angle Method

The course of three-point method is severely curved. If a missile leads an angle relative to the line of sight from the guidance station to a target, then the course of the missile becomes less curved than three-point method. If the $f(t) = -\dfrac{1}{\Delta \dot{r}} \dot{\varepsilon}_T$ or $f(t) = -\dfrac{1}{\Delta \dot{r}} \dot{\beta}_T \cos \varepsilon_T$, where $\Delta r = R_T - R_M$ is missile-target relative distance, then the lead angle guidance equations are

$$\varepsilon_M = \varepsilon_T - \frac{\dot{\varepsilon}_T}{\Delta \dot{r}} \Delta r \qquad (7.2-23)$$

$$\beta_M = \beta_T - \frac{\dot{\beta}_T \cos \varepsilon_T}{\Delta \dot{r}} \Delta r \qquad (7.2-24)$$

Especially, half lead angle method is widely used in remote control. The half lead angle guidance equations are

$$\varepsilon_M = \varepsilon_T - \frac{\dot{\varepsilon}_T}{2\Delta \dot{r}} \Delta r \qquad (7.2-25)$$

$$\beta_M = \beta_T - \frac{\dot{\beta}_T \cos \varepsilon_T}{2\Delta \dot{r}} \Delta r \qquad (7.2-26)$$

7.3 Homing Guidance Relative Motion Equations

Figure 7.3-1 illustrates the missile-target geometrical relation in space.

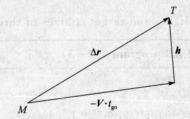

Figure 7.3-1 Missile-target geometrical relation in space

From figure 7.3-1 it can be seen that instantaneous miss distance vector $\boldsymbol{h}$ is

$$\boldsymbol{h} = \Delta \boldsymbol{r} + t_{go} \boldsymbol{V} \qquad (7.3-1)$$

where

$\Delta \boldsymbol{r} = \boldsymbol{R}_T - \boldsymbol{R}_M$, the missile-target relative distance vector,

$\boldsymbol{V} = \boldsymbol{V}_T - \boldsymbol{V}_M = \dfrac{d\Delta \boldsymbol{r}}{dt}$, the missile-target relative velocity vector,

t_{go} = the time-to-go (time-to-go is defined as the time remaining to intercept; mathematically, $t_{go} = t_f - t$, where t_f is the final time and t is the present time).

In addition, there are still these relations as follows:

$$|\Delta \boldsymbol{r}| = \Delta r$$

$$|\boldsymbol{V}|=V$$
$$\boldsymbol{a}=\frac{\mathrm{d}\boldsymbol{V}}{\mathrm{d}t}=\boldsymbol{a}_T-\boldsymbol{a}_M$$

where

$\boldsymbol{a}=$ the missile-target relative acceleration vector,

$\boldsymbol{a}_T=$ the target acceleration vector,

$\boldsymbol{a}_M=$ the missile acceleration vector.

Taking derivative of the equation (7.3-1) yields

$$\dot{\boldsymbol{h}}=\frac{\mathrm{d}\Delta\boldsymbol{r}}{\mathrm{d}t}-\boldsymbol{V}+t_{\mathrm{go}}\frac{\mathrm{d}\boldsymbol{V}}{\mathrm{d}t}=\boldsymbol{V}-\boldsymbol{V}+t_{\mathrm{go}}\boldsymbol{a}=t_{\mathrm{go}}\boldsymbol{a}$$

i.e.

$$\dot{\boldsymbol{h}}=t_{\mathrm{go}}(\boldsymbol{a}_T-\boldsymbol{a}_M) \qquad (7.3-2)$$

The angular velocity $\boldsymbol{\omega}_{\mathrm{LOS}}$ perpendicular to line of sight $\Delta\boldsymbol{r}$ is

$$\boldsymbol{\omega}_{\mathrm{LOS}}=\frac{1}{\Delta r}(\Delta\boldsymbol{r}^{\circ}\times\boldsymbol{V}) \qquad (7.3-3)$$

where

$\Delta\boldsymbol{r}^{\circ}=\frac{\Delta\boldsymbol{r}}{|\Delta\boldsymbol{r}|}=\frac{\Delta\boldsymbol{r}}{\Delta r}$, the unit vector of relative distance.

After both sides of equation (7.3-3) are multiplied by $\Delta\boldsymbol{r}^{\circ}$,

$$\boldsymbol{\omega}_{\mathrm{LOS}}\times\Delta\boldsymbol{r}^{\circ}=\frac{1}{\Delta r}(\Delta\boldsymbol{r}^{\circ}\times\boldsymbol{V})\times\Delta\boldsymbol{r}^{\circ}=$$
$$\frac{1}{\Delta r}\boldsymbol{V}(\Delta\boldsymbol{r}^{\circ}\Delta\boldsymbol{r}^{\circ})-\frac{1}{\Delta r}\Delta\boldsymbol{r}^{\circ}(\boldsymbol{V}\Delta\boldsymbol{r}^{\circ})=\frac{1}{\Delta r}[\boldsymbol{V}-\Delta\boldsymbol{r}^{\circ}(\boldsymbol{V}\Delta\boldsymbol{r}^{\circ})]$$

In view of $\Delta\boldsymbol{r}=-\Delta\boldsymbol{r}^{\circ}t_{\mathrm{go}}(\Delta\boldsymbol{r}^{\circ}\boldsymbol{V})$, the above equation becomes

$$\boldsymbol{\omega}_{\mathrm{LOS}}\times\Delta\boldsymbol{r}^{\circ}=\frac{1}{\Delta r}\left(\boldsymbol{V}+\frac{\Delta\boldsymbol{r}}{t_{\mathrm{go}}}\right)$$

After both sides of above equation are multiplied by $\Delta r t_{\mathrm{go}}$,

$$\boldsymbol{h}=\boldsymbol{V}t_{\mathrm{go}}+\Delta\boldsymbol{r}=\Delta r t_{\mathrm{go}}(\boldsymbol{\omega}_{\mathrm{LOS}}\times\Delta\boldsymbol{r}^{\circ})=Vt_{\mathrm{go}}^{2}(\boldsymbol{\omega}_{\mathrm{LOS}}\times\Delta\boldsymbol{r}^{\circ}) \qquad (7.3-4)$$

The equations (7.3-2), (7.3-3) and (7.3-4) are homing guidance motion equation vector forms in inertial coordinate system.

In order to obtain more intuitionistic expressions, suppose that the missile-target lies in the same plumb plane as shown in Figure 7.3-2. This case will be discussed.

CHAPTER 7 Guidance Laws

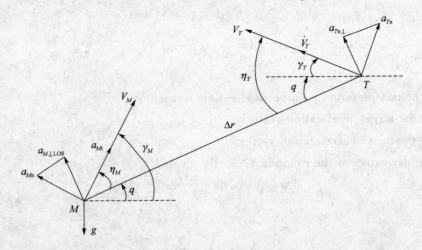

Figure 7.3-2 Missile-target geometrical relation in plumb plane

From the equation (7.3-2),

$$\dot{h} = t_{go}(a_{Tn\perp} - a_{Mn\perp}) \tag{7.3-5}$$

$$h = V t_{go}^2 \omega_{LOS} \tag{7.3-6}$$

$$\Delta \dot{r} = -V_T \cos(\gamma_T + q) - V_M \cos(\gamma_M - q) \tag{7.3-7}$$

$$\omega_{LOS} = \dot{q} = \frac{V_T \sin(\gamma_T + q) - V_M \sin(\gamma_M - q)}{\Delta r} \tag{7.3-8}$$

$$\eta_M = \gamma_M - q \tag{7.3-9}$$

$$\eta_T = \gamma_T + q \tag{7.3-10}$$

From the equation (7.3-8),

$$\Delta r \omega_{LOS} = V_T \sin(\gamma_T + q) - V_M \sin(\gamma_M - q)$$

Taking derivative of the above equation yields

$$\Delta r \dot{\omega}_{LOS} + \Delta \dot{r} \omega_{LOS} = \dot{V}_T \sin(\gamma_T + q) + V_T \cos(\gamma_T + q)(\dot{\gamma}_T + \dot{q}) -$$

$$\dot{V}_M \sin(\gamma_M - q) - V_M \cos(\gamma_M - q)(\dot{\gamma}_M - \dot{q}) =$$

$$V_T \cos(\gamma_T + q)\dot{q} + V_M \cos(\gamma_M - q)\dot{q} +$$

$$\dot{V}_T \sin(\gamma_T + q) + V_T \cos(\gamma_T + q)\dot{\gamma}_T -$$

$$\dot{V}_M \sin(\gamma_M - q) - V_M \cos(\gamma_M - q)\dot{\gamma}_M =$$

$$-\Delta \dot{r} \omega_{LOS} + \dot{V}_T \sin(\gamma_T + q) + V_T \dot{\gamma}_T \cos(\gamma_T + q) -$$

$$\dot{V}_M \sin(\gamma_M - q) - V_M \dot{\gamma}_M \cos(\gamma_M - q)$$

Thus

$$\Delta r \dot{\omega}_{LOS} + 2\Delta \dot{r}\omega_{LOS} = \dot{V}_T \sin(\gamma_T + q) + V_T \dot{\gamma}_T \cos(\gamma_T + q) -$$
$$\dot{V}_M \sin(\gamma_M - q) - V_M \dot{\gamma}_M \cos(\gamma_M - q) =$$
$$a_{Tn\perp} - a_{Mn\perp} \qquad (7.3-11)$$

where

$a_{Tn\perp} = \dot{V}_T \sin(\gamma_T + q) + V_T \dot{\gamma}_T \cos(\gamma_T + q)$, the target acceleration perpendicular to line of sight,

$a_{Mn\perp} = \dot{V}_M \sin(\gamma_M - q) + V_M \dot{\gamma}_M \cos(\gamma_M - q)$, the missile acceleration perpendicular to line of sight.

The equation (7.3 - 11) is well-known kinematical equation of homing guidance in a plane.

In view of the target tangent acceleration $a_{Tt} = \dot{V}_T$ and normal acceleration $a_{Tn} = V_T \dot{\gamma}_T$, thus

$$a_{Tn\perp} = a_{Tt} \sin(\gamma_T + q) + a_{Tn} \cos(\gamma_T + q) \qquad (7.3-12)$$

If consider controlled acceleration and gravity acceleration distinctively, then

$$\dot{V}_M = a_{Mt} - g\sin\gamma_M, \quad V_M \dot{\gamma}_M = a_{Mn} - g\cos\gamma_M \qquad (7.3-13)$$

where, a_{Mt} and a_{Mn} are respectively tangent and normal accelerations, which are produced by aerodynamic force and counterforce.

Therefore

$$a_{Mn\perp} = a_{Mt}\sin(\gamma_M - q) + a_{Mn}\cos(\gamma_M - q) - g\cos q =$$
$$a_{Mt}\sin(\gamma_M - q) + a_{M\perp LOS} - g\cos q \qquad (7.3-14)$$

where, $a_{M\perp LOS} = a_{Mn}\cos(\gamma_M - q)$ expresses the missile acceleration, perpendicular to line of sight, produced by control command.

7.4 Pursuit Method

A pure pursuit course is a course in which the missile velocity vector is always directed toward the instantaneous target position. Lead angle η_M of missile velocity vector in Figure 7.3 - 2 is equal to zero. The equations (7.3-7), (7.3-8), (7.3-9) and (7.3-10) become

$$\Delta \dot{r} = -V_T \cos(\gamma_T + q) - V_M \qquad (7.4-1)$$

$$\omega_{LOS} = \dot{q} = \frac{V_T \sin(\gamma_T + q)}{\Delta r} \qquad (7.4-2)$$

$$\eta_M = \gamma_M - q = 0 \qquad (7.4-3)$$
$$\eta_T = \gamma_T + q \qquad (7.4-4)$$

They are the relative motion equations of pursuit method. If V_M, V_T and γ_T are known, the equations contain three unknown parameters, Δr, q and η_T. If initial values Δr_0, q_0 and η_{T0} are given, the solutions may be obtained by digital integral.

Figure 7.4-1 shows guidance trajectory of pursuit method. Supposing that the target and missile move in the same plumb plane, absolute trajectory is plotted in the Earth surface coordinate system Oxz in Figure 7.4-1 (a). The target flies at constant speed along straight line course. M_0, M_1, M_2, ... and T_0, T_1, T_2, ... respectively correspond to the missile and target locations at the same moment. The tangent directions of the missile course coincide with the line of sight. Referring to the target, the trajectory of the missile is plotted in Figure 7.4-1 (b). That is the trajectory, which one observes from the target. Although the missile velocity always points at the target, the relative velocity V_R always falls behind the line of sight. Therefore, the course of the missile becomes more and more curved, and requires greater normal load factor for the missile.

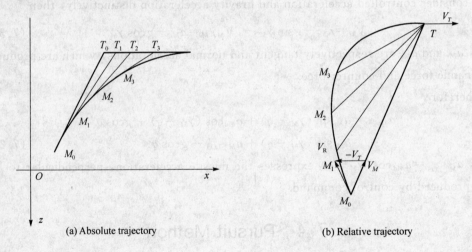

(a) Absolute trajectory   (b) Relative trajectory

Figure 7.4-1 Guidance trajectory of pursuit method

The pursuit method has strict limitation to the ratio of missile speed to target speed, $p = V_M/V_T$. Necessary condition of impacting target is

$$1 < p = V_M/V_T \leqslant 2 \qquad (7.4-5)$$

At present, the pursuit method is seldom used.

7.5 Constant-bearing Guidance

A constant-bearing course is a course in which the line-of-sight from the missile to the target maintains a constant direction in space. The lines of sight keep parallel, and the corresponding guidance equation is

$$\dot{q}=0 \tag{7.5-1}$$

Thus, the equation (7.3-8) becomes

$$V_T \sin(\gamma_T + q) = V_M \sin(\gamma_M - q) \tag{7.5-2}$$

i.e.

$$V_T \sin \eta_T = V_M \sin \eta_M \tag{7.5-3}$$

The equation (7.5-2) shows that the component of missile velocity perpendicular to line of sight is equal to that of target velocity perpendicular to line of sight. Relative velocity V_R always points at the target as shown in Figure 7.5-1.

If a target flies along straight line course, then the course of a missile is a straight line. Although the course of a missile is less curved, it is difficult to maintain the equation (7.5-3) relation in guidance process. Thus the guidance method is difficultly implemented.

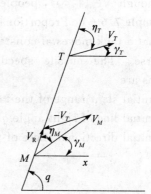

Figure 7.5-1 Geometrical relation of constant-bearing guidance

7.6 Proportional Navigation

A proportional-navigation course is a course in which the changing rate of the missile heading is directly proportional to the rate of rotation of the line-of-sight from the missile to the target. Thus, the basic differential equation for this case is given by

$$\dot{\gamma}_M = N\dot{q} = N\omega_{\text{LOS}} \tag{7.6-1}$$

where N is the navigation constant.

Taking derivative of $\gamma_M = \eta_M + q$ with respect to t yields

$$\dot{\gamma}_M = \dot{\eta}_M + \dot{q} \tag{7.6-2}$$

Substituting equation (7.6-1) into equation (7.6-2) yields

$$\dot{\eta}_M = (N-1)\dot{q} \tag{7.6-3}$$

From equation (7.6-3), it can be seen that if $N=1$, then $\dot{\eta}_M=0$, $\eta=\eta_0=$ constant, corresponding to constant lead angle guidance. Specially, $\eta=0$ corresponds to the pursuit guidance. If $N\to\infty$, then $\frac{dq}{dt}\to 0$, i.e. $q=q_0=$ constant, corresponds to constant-bearing guidance. Therefore pursuit method and constant-bearing guidance are two special situations of proportional navigation. In other words, the trajectory of proportional navigation is located between trajectories of the two methods. Thus the trajectory characteristics of proportional navigation lie between pursuit method and constant-bearing guidance as seen in Figure 7.6-1.

Although $N\in(1,\infty)$, people often take on the value of $2<N<6$.

Example 7.6-1: Proportional navigation guidance of anti-tank missile. Figure 7.6-2 shows that a tank moves at constant velocity $V_T=12$m/s along a non-maneuvering straight-line course. The missile speed V_M is constant, $V_M=120$m/s. Initial intercepting conditions are
- initial slant range of the target from the missile $\Delta r_0 = 3\,000$ m,
- initial line of sight angle $q=70°$,
- initial directional angle of the missile velocity $\psi_{M0}=72°$.

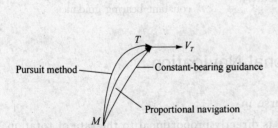

Figure 7.6-1 Comparison of the three guidance laws

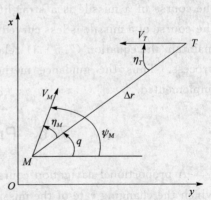

Figure 7.6-2 Proportional navigation guidance of an anti-tank missile

The motion equations are written as
$$\left.\begin{array}{l} \Delta\dot{r} = -V_T\cos q - V_M\cos \eta_M \\ \dot{q} = V_T\sin q - V_M\sin \eta_M \\ \psi_M = q + \eta_M \\ \dot{\psi}_M = N\dot{q} \end{array}\right\} \tag{7.6-4}$$

7.6 Proportional Navigation

Integrating the forth equation of the equations (7.6-4) yields

$$\psi_M = \psi_{M0} + N(q - q_0) \tag{7.6-5}$$

Substituting the equation (7.6-5) into the third equation of the equations (7.6-4) yields

$$\eta_M = -Nq_0 + \psi_{M0} + (N-1)q \tag{7.6-6}$$

Substituting the equation (7.6-6) into the first equation and second equation of the equations (7.6-4) yields

$$\left.\begin{array}{l} \Delta \dot{r} = -V_T \cos q - V_M \cos[-Nq_0 + \psi_{M0} + (N-1)q] \\ \dot{q} = V_T \sin q - V_M \sin[-Nq_0 + \psi_{M0} + (N-1)q] \end{array}\right\} \tag{7.6-7}$$

Take the navigation constant $N=4$. In a similar manner, the Runge-Kutta integrator ode45 of MATLAB is used to solve the differential equations (7.6-7). The M file is given by MATLAB program 7.6-1.

MATLAB program 7.6-1

% Proportional navigation guidance for an anti-tank missile

```
function xdot=pnm(t,x)
% DeltR=x(1);        ! Missile slant range
% q=x(2);            ! Missile line of sight angle in radian
% xT=x(3);           ! Target x axis coordinate
% xT=x(4);           ! Target y axis coordinate

VM=120;              % Missile speed
VT=12;               % Target speed
phi0=72/57.29578;    % Initial angle of Missile velocity
N=4;                 % Navigation constant
q0=70/57.29578;      % Initial angle of LOS

xdot(1)=-VT*cos(x(2))-VM*cos(-N*q0+phi0+(N-1)*x(2));
xdot(2)=(VT*sin(x(2))-VM*sin(-N*q0+phi0+(N-1)*x(2)))/x(1);
xdot(3)=0;
xdot(4)=-12;
xdot=xdot';
```

In order to plot missile and target courses, the following equations are needed:

$$\left.\begin{array}{l} x_M = x_T - \Delta r \cos q \\ y_M = y_T - \Delta r \sin q \end{array}\right\} \tag{7.6-8}$$

157

CHAPTER 7 Guidance Laws

Entering MATALB program 7.6-2 in command window of MATALB yields the missile and target's courses as shown in Figure 7.6-3.

MATLAB program 7.6-2

```
>> [t,x]=ode45('pnm',[0 25],[3000 70/57.29578 3000*sin(70/57.29578) 3000*cos(70/57.29578)]);
>> [m,n]=size(x(:,:));
>> for i=1:m
       xt(i)=x(i,3);
       yt(i)=x(i,4);
       xm(i)=xt(i)-x(i,1)*sin(x(i,2));
       ym(i)=yt(i)-x(i,1)*cos(x(i,2));
   end
>> plot(ym,xm,'black',yt,xt,'black')
>> xlabel('y (m)');
>> ylabel('x (m)');
```

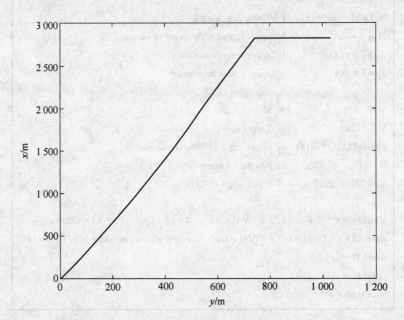

Figure 7.6-3 Missile and target courses of proportional navigation guidance

Now estimate the impact time. Since the impact point lies on the course of target, the corresponding coordinate of the missile in x axis at impact point is

$$x_{Mf} = R_{T0} \sin q_{T0} = 3\,000 \text{ m} \times 0.939\,7 = 2\,819.1 \text{ m}$$

The interpolation is used to evaluate the impact time t_f. Entering MATLAB program 7.6-3 produces t_f.

MATLAB program 7.6-3

```
>> x0=xm';
>> y0=ym';
>> ymf=interp1(x0,y0,2819.1,'cubic');
>> tf=(3000*cos(70/57.29578)-ymf)/12
tf = 24.2826
```

Thus the total elapsed time to impact the target from launch, t_f is about $t_f = 24.28$ s.

Below, using the well-known kinematic equation (7.3-11) of the homing guidance, the proportional navigation is discussed deeply. Substituting $\dot{\gamma}_M = N\dot{q} = N\omega_{LOS}$ into equation (7.3-11) yields

$$\Delta r \dot{\omega}_{LOS} + [2\Delta \dot{r} + V_M N \cos(\gamma_M - q)]\omega_{LOS} =$$
$$\dot{V}_T \sin(\gamma_T + q) + V_T \dot{\gamma}_T \cos(\gamma_T + q) - \dot{V}_M \sin(\gamma_M - q)$$

or

$$\dot{\omega}_{LOS} + \frac{1}{\Delta r}[2\Delta \dot{r} + V_M N \cos(\gamma_M - q)]\omega_{LOS} =$$
$$\frac{1}{\Delta r}[\dot{V}_T \sin(\gamma_T + q) + V_T \dot{\gamma}_T \cos(\gamma_T + q) - \dot{V}_M \sin(\gamma_M - q)] \quad (7.6-9)$$

Let right side of equation (7.6-9) be equal to zero, and then obtains a homogeneous differential equation about ω_{LOS}. In order for ω_{LOS} to converge, there is the following inequality.

$$2\Delta \dot{r} + V_M N \cos(\gamma_M - q) > 0 \quad (7.6-10)$$

From the inequality (7.6-10), the navigation ratio N satisfies

$$N > [-2\Delta \dot{r}/V_M \cos(\gamma_M - q)] \quad \text{for} \quad \cos(\gamma_M - q) > 0 \quad (7.6-11)$$

Substituting $\Delta \dot{r}$ from equation (7.3-7) into inequality (7.6-11) yields

$$N > 2 \times \left[1 + \frac{\cos(\gamma_T + q)}{p \cos(\gamma_M - q)}\right] \quad (7.6-12)$$

where $p = V_M/V_T$. Now let

$$N = -N'\Delta \dot{r}/V_M \cos(\gamma_M - q) \quad (7.6-13)$$

CHAPTER 7 Guidance Laws

Then

$$N' = -N[V_M \cos(\gamma_M - q)/\Delta\dot{r}] \qquad (7.6-14)$$

where $N'(N'>2)$ is commonly called the effective navigation ratio, and $-\Delta\dot{r}$ is the missile-target closing velocity (i. e. $-\Delta\dot{r} = V_C$).

Since the missile velocity vector can not be controlled directly, the missile normal acceleration a_n is defined as

$$a_n = V_M \dot{\gamma}_M \qquad (7.6-15)$$

where $\dot{\gamma}_M$ is the missile's turning rate. Substituting $\dot{\gamma}_M = N\dot{q}$ into equation (7.6-15) yields

$$a_n = V_M N \dot{q} \qquad (7.6-16)$$

Substituting the equation (7.6-13) into the equation (7.6-16) yields

$$a_n = \{-N'\Delta\dot{r}/[V_M \cos(\gamma_M - q)]\} V_M \dot{q} =$$

$$N' \frac{V_C}{\cos(\gamma_M - q)} \dot{q} = N' \frac{V_C}{\cos(\gamma_M - q)} \omega_{LOS} \qquad (7.6-17)$$

Equation (7.6-17) is the well-known general classical proportional navigation guidance equation. The missile's lead angle $(\gamma_M - q)$ can be measured by a missile's seeker. If missile's lead angle $(\gamma_M - q)$ is very small, then

$$a_n = N' V_C \dot{q} = N' V_C \omega_{LOS} \qquad (7.6-18)$$

Equation (7.6-18) is widely used in homing guidance.

From the equations (7.3-11) and (7.3-14),

$$\Delta r \dot{\omega}_{LOS} + 2\Delta\dot{r}\omega_{LOS} = \dot{V}_T \sin(\gamma_T + q) + V_T \dot{\gamma}_T \cos(\gamma_T + q) -$$

$$a_{Mt} \sin(\gamma_M - q) + g\cos q - a_{M\perp LOS} \qquad (7.6-19)$$

The solution of the differential equation will be discussed as follows.

1. Proportional navigation without any lead quantity

For proportional navigation guidance, substituting $a_{M\perp LOS} = NV_C \omega_{LOS}$ into equation (7.6-19) yields

$$\omega_{LOS} = \omega_{LOS0} \left(\frac{\Delta r}{\Delta r_0}\right)^{N-2} -$$

$$\frac{\dot{V}_T \sin(\gamma_T + q) + V_T \dot{\gamma}_T \cos(\gamma_T + q) - a_{Mt} \sin(\gamma_M - q) + g\cos q}{(N-2)\Delta\dot{r}} \left[1 - \left(\frac{\Delta r}{\Delta r_0}\right)^{N-2}\right] \qquad (7.6-20)$$

Thus

$$a_{M \perp \text{LOS}} = NV_C \omega_{\text{LOS0}} \left(\frac{\Delta r}{\Delta r_0}\right)^{N-2} +$$

$$N \frac{\dot{V}_T \sin(\gamma_T + q) + V_T \dot{\gamma}_T \cos(\gamma_T + q) - a_{Mt} \sin(\gamma_M - q) + g\cos q}{(N-2)} \left[1 - \left(\frac{\Delta r}{\Delta r_0}\right)^{N-2}\right]$$

(7.6-21)

For the proportional navigation guidance, target maneuvering, missile accelerating and gravity belong to equivalent disturbances and cause needed acceleration to increase. At impact point needed acceleration reaches its maximum:

$$a_{M \perp \text{LOS}(\Delta r = 0)} = N \frac{\dot{V}_T \sin(\gamma_T + q) + V_T \dot{\gamma}_T \cos(\gamma_T + q) - a_{Mt} \sin(\gamma_M - q) + g\cos q}{(N-2)}$$

(7.6-22)

These disturbances cause the missile's course to be more curved.

2. Proportional navigation with lead quantity

The guidance equation is defined as

$$a_n = NV_C(\omega_{\text{LOS}} + \omega_{Mt} + \omega_g + \omega_T) \quad (7.6-23)$$

where

a_n = needed acceleration of missile,

N = navigation coefficient,

V_C = closing velocity,

ω_{LOS} = line-of-sight angular velocity,

ω_{Mt} = compensation missile tangent acceleration component,

ω_g = compensation missile gravity component,

ω_T = compensation target maneuvering component.

For proportional navigation with lead quantity, let

$$a_{M \perp \text{LOS}} = NV_C(\omega_{\text{LOS}} + \omega_{Mt} + \omega_g + \omega_T) \quad (7.6-24)$$

Substituting the equation (7.6-24) into the equation (7.6-19) yields

$$\dot{\omega}_{\text{LOS}} + \frac{-\Delta \dot{r}}{\Delta r}(N-2)\omega_{\text{LOS}} =$$

$$\frac{\dot{V}_T \sin(\gamma_T + q) + V_T \dot{\gamma}_T \cos(\gamma_T + q) - a_{Mt} \sin(\gamma_M - q) + g\cos q - NV_C \omega_{Mt} - NV_C \omega_g - NV_C \omega_T}{\Delta r}$$

(7.6-25)

The solution of the differential equation (7.6-25) is

$$\omega_{LOS} = \omega_{LOS0}\left(\frac{\Delta r}{\Delta r_0}\right)^{N-2} - A\left[1 - \left(\frac{\Delta r}{\Delta r_0}\right)^{N-2}\right] \qquad (7.6-26)$$

where

$$A = \frac{\dot{V}_T \sin(\gamma_T + q) + V_T \dot{\gamma}_T \cos(\gamma_T + q) - a_{Mt}\sin(\gamma_M - q) + g\cos q - NV_C\omega_{Mt} - NV_C\omega_g - NV_C\omega_T}{(N-2)\Delta \dot{r}},$$

ω_{LOS0} = initial value of line-of-sight angular velocity,

Δr_0 = initial value of missile-target relative distance.

Substituting the equation (7.6-26) into the equation (7.6-24) yields

$$a_{M\perp LOS} = NV_C\omega_{LOS0}\left(\frac{\Delta r}{\Delta r_0}\right)^{N-2} + NV_C A\left(\frac{\Delta r}{\Delta r_0}\right)^{N-2} - NV_C(A - \omega_{Mt} - \omega_g - \omega_T) \qquad (7.6-27)$$

The equation (7.6-27) shows that at initial time ($\Delta r = \Delta r_0$) the needed acceleration is $a_{M\perp LOS} = NV_C(\omega_{LOS0} + \omega_{Mt} + \omega_g + \omega_T)$. At impact point ($\Delta r = 0$), the needed acceleration is

$$a_{M\perp LOS(\Delta r = 0)} = -NV_C A + NV_C(\omega_{Mt} + \omega_g + \omega_T) \qquad (7.6-28)$$

If the needed acceleration at impact point ($\Delta r = 0$) is required to be zero, i.e. $a_{M\perp LOS(\Delta r=0)} = 0$, then

$$\omega_{Mt} + \omega_g + \omega_T = A =$$
$$\frac{\dot{V}_T \sin(\gamma_T + q) + V_T\dot{\gamma}_T\cos(\gamma_T + q) - \dot{V}_M\sin(\gamma_M - q) + g\cos q - NV_C\omega_{Mt} - NV_C\omega_g - NV_C\omega_T}{(N-2)\Delta \dot{r}}$$

$$(7.6-29)$$

If choosing

$$\omega_{Mt} = -\frac{\dot{V}_M \sin(\gamma_M - q)}{2V_C} \qquad (7.6-30)$$

$$\omega_g = \frac{g\cos q}{2V_C} \qquad (7.6-31)$$

$$\omega_T = \frac{\dot{V}_T\sin(\gamma_T + q) + a_{Tn}\cos(\gamma_T + q)}{2V_C} \qquad (7.6-32)$$

then the equation (7.6-29) is satisfied. This means that if these compensation terms are taken according to the equations (7.6-30), (7.6-31) and (7.6-32), the needed acceleration at impact point ($\Delta r = 0$) is zero.

3. Proportional navigation with partial lead quantity

Actually, it is difficult to measure target maneuvering parameters, so people normally

7.6 Proportional Navigation

only compensate missile tangent acceleration and missile gravity. Let

$$a_{M\perp LOS} = NV_C(\omega_{LOS} + \omega_{Mt} + \omega_g) \tag{7.6-33}$$

Substituting the equation (7.6-33) into (7.6-19) yields

$$\dot{\omega}_{LOS} + \frac{-\Delta\dot{r}}{\Delta r}(N-2)\omega_{LOS} =$$

$$\frac{\dot{V}_T\sin(\gamma_T+q) + V_T\dot{\gamma}_T\cos(\gamma_T+q) - a_{Mt}\sin(\gamma_M-q) + g\cos q - NV_C\omega_{Mt} - NV_C\omega_g}{\Delta r}$$

$$\tag{7.6-34}$$

The solution of the differential equation (7.6-34) is

$$\omega_{LOS} = \omega_{LOS0}\left(\frac{\Delta r}{\Delta r_0}\right)^{N-2} -$$

$$\frac{\dot{V}_T\sin(\gamma_T+q) + V_T\dot{\gamma}_T\cos(\gamma_T+q) - a_{Mt}\sin(\gamma_M-q) + g\cos q - NV_C\omega_{Mt} - NV_C\omega_g}{(N-2)\Delta\dot{r}}\left[1-\left(\frac{\Delta r}{\Delta r_0}\right)^{N-2}\right]$$

$$\tag{7.6-35}$$

Substituting the equation (7.6-35) into (7.6-33) yields

$$a_{M\perp LOS} = NV_C(\omega_{LOS} + \omega_{Mt} + \omega_g) = NV_C\omega_{LOS0}\left(\frac{\Delta r}{\Delta r_0}\right)^{N-2} -$$

$$NV_C\frac{\dot{V}_T\sin(\gamma_T+q) + V_T\dot{\gamma}_T\cos(\gamma_T+q) - a_{Mt}\sin(\gamma_M-q) + g\cos q - NV_C\omega_{Mt} - NV_C\omega_g}{(N-2)\Delta\dot{r}} \cdot$$

$$\left[1-\left(\frac{\Delta r}{\Delta r_0}\right)^{N-2}\right] + NV_C\omega_{Mt} + NV_C\omega_g \tag{7.6-36}$$

Let

$$\omega_{Mt} = -\frac{\dot{V}_M\sin(\gamma_M-q)}{2V_C} \tag{7.6-37}$$

$$\omega_g = \frac{g\cos q}{2V_C} \tag{7.6-38}$$

Then

$$a_{M\perp LOS} = a_n = NV_C\left[\omega_{LOS0} - \frac{a_{Mt}\sin(\gamma_M-q)}{2V_C} + \frac{g\cos q}{2V_C}\right]\left(\frac{\Delta r}{\Delta r_0}\right)^{N-2} -$$

$$NV_C\frac{\dot{V}_T\sin(\gamma_T+q) + V_T\dot{\gamma}_T\cos(\gamma_T+q)}{(N-2)\Delta\dot{r}}\left[1-\left(\frac{\Delta r}{\Delta r_0}\right)^{N-2}\right] =$$

$$NV_C\left[\omega_{LOS0} - \frac{a_{Mt}\sin(\gamma_M-q)}{2V_C} + \frac{g\cos q}{2V_C}\right]\left(\frac{\Delta r}{\Delta r_0}\right)^{N-2} +$$

$$N\frac{\dot{V}_T\sin(\gamma_T+q) + V_T\dot{\gamma}_T\cos(\gamma_T+q)}{(N-2)}\left[1-\left(\frac{\Delta r}{\Delta r_0}\right)^{N-2}\right] \tag{7.6-39}$$

CHAPTER 7　Guidance Laws

References

[1] George M Siouris. Missile Guidance and Control. New York:Spring-Verlag Inc. ,2004.
[2] 斯维特洛夫ＢＴ,戈卢别夫ＮＣ,等.防空导弹设计.北京:中国宇航出版社,2004.
[3] 娄寿春.导弹制导技术.北京:中国宇航出版社,1989.
[4] 钱杏芳,林瑞雄,赵亚男.导弹飞行力学.北京:北京理工大学出版社,2000.

CHAPTER 8
Autopilot Design

8.1 Introduction

The previous chapters have built vehicle mathematical models, and treated flight control structures for performing flight control design and simulation. This chapter will treat control law design of vehicle based on time-domain, frequency-domain, and root-locus methods of classical control theory by MATLAB.

In time-domain method, by entering a typical signal to control system, one observes the system response with respect to the input to rate the system performance. In design, there are step, ramp, and impulse signals in common use. The characteristic parameters of step response, such as rise time, peak time, settling time, and overshoot, normally act as time-domain indexes.

The root locus method gives a graphic picture of the movement of the poles of closed-loop system with the variation of one of the system parameters. In terms of the root locus diagram, on one hand, if any of the roots of the characteristic equation of the closed-loop system lies in the right half plane the system is unstable. On the other hand, if all the roots lie in the left half plane the system is stable. The farther the roots in the left half plane are from the imaginary axis, the faster the response decays. All poles lying along a line perpendicular to the real axis have the same time to half amplitude. Poles lying along the same horizontal line have the same damped frequency and period. Poles lying along a radial line through the origin have the same damping ratio. Poles lying along the same circular arc around the origin have the same undamped natural frequency. Complex roots lying on the imaginary axis yield constant amplitude oscillation. Repeated roots on the imaginary axis result in unstable behavior. In many cases, it is not possible to satisfy all the performance specifications using a single parameter such as the system gain. This requires the designer to add poles and zeros, and reshape root locus as desired so that the root locus passes through desired closed-loop poles on s-plane. These poles and zeros form a transfer function of a compensator. For example, there are lead, lag, and lag-lead compensators and PID controller. The compensator can compensate for deficit performance of the original system. Note that the addition of a compensator generally increases the order of the system. The designer must check the performance specifications of the final system to be sure they are satisfactory. If not, repeat the design process by modifying the pole-zero location of the compensator until a satisfactory result is obtained.

There are basically two approaches in the frequency-domain design. One is the polar plot approach. The other is the Bode diagram approach. In general, it is convenient to use Bode diagram for design purposes. The gain margin and phase margin are used as the frequency domain performance specifications. To achieve satisfactory performance, the phase margin should be between 30° and 60°, and the gain margin should be greater than 6 dB. A common approach to the Bode diagram is that the open-loop gain should be adjusted first so that the requirement on the steady-state accuracy can be met. Then the magnitude and phase curves of the uncompensated open loop (only the open-loop gain just adjusted) are plotted. If the specifications on the phase and gain margins are not satisfied, then a designer has to choose a suitable compensator to reshape the Bode diagram of the open-loop function. When there are mutually conflicting requirements, a compromise is still needed. After the open loop has been designed by frequency-response method, the closed-loop poles and zeros can be determined. The transient-response must be checked to see whether the designed system satisfies the requirements in the time domain. If it does not, the compensator must be improved and the analysis should be repeated until a satisfactory result is obtained.

In essence, the design course of the classical control theory is based on the trial-and-error. In design course, the designer often adjusts the parameters of a compensator, and checks the system performance by analysis with each adjustment of the parameters. Only if all performances are met, the trial-and-error process is over. It is noted that in designing control systems by the root-locus or frequency-response methods, the final result is not unique.

There are uncertainties and changes in the CG due to propellant consumption and manufacturing tolerances. The propellant consumption causes mass and inertia changes. The aerodynamic derivatives change with dynamic pressure and height variations. These uncertainties and variations result in uncertainties and variations of vehicle model. This complicates the design of the flight control system and limits system performances. The control system has to be adapted (gain-scheduled, or self-adaptive) in flight, allow for the wide variations in vehicle dynamics over the large flight envelop. The flight controller must be adjusted according to flight condition. A well designed autopilot can tolerate considerable changes in environments. The adjustment process is called gain scheduling. One simple form is that the control parameters change with dynamic pressure and/or Mach number.

8.2 Autopilot of Roll Channel

Consider a ground-to-air missile whose roll position needs to be stabilized. The aerodynamic parameters change in a wide range due to the variability in speed and altitude. Table 8.2 - 1 shows the variability of the roll derivatives, aerodynamic gain, and time constant of a missile. The roll moment of inertia is $J_x = 38.76$ kg·m².

Table 8.2 - 1 Roll derivatives, aerodynamic gain and time constant

t/s	$-L_p$	$-L_{\delta_a}$	$-K_a = -L_{\delta_a}/L_p$	$T_{Mx} = -1/L_p$
0	6.912	621	−89.88	0.145
8	5.373	493	−91.67	0.186
16	3.695	344	−93.06	0.271
22	2.300	218	−94.84	0.435
30	0.900	92	−102.30	1.114
32.5	0.637	66	−104.40	1.571

In remote control guidance process for the ground to air missile, the control commands from the guidance station are commonly decomposed onto two mutually orthogonal planes. In order to correctly implement control commands, the roll angle needs to be stabilized. It is essential to maintain roll angle to be zero or as small as possible. In actual system, there exist disturbing torques. For the sake of cancelling disturbance torques, people design a stabilized loop for roll channel. Figure 8.2 - 1 shows the roll position control loop.

It is estimated that the largest rolling moment will occur at $t = 16$ s due to unequal incidence in pitch and yaw, and will have a maximum value 1 000 N·m. If the maximum missile roll angle permissible is 1/20 rad, then the stiffness of the loop must be not less than 1 000/(1/20) N·m/rad = 20 000 N·m/rad. This means that in order to balance this

CHAPTER 8 Autopilot Design

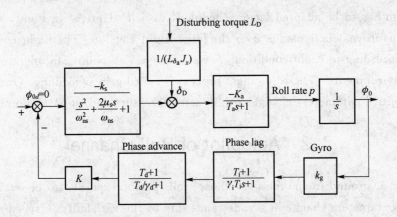

Figure 8.2-1 Roll position control loop

disturbance torque one has to use $1\,000/(344\times38.76)$ rad aileron, and this is approximately 4.3°. In order to achieve a negative feedback system, the actual servo steady gain $-k_s$ has to be negative. Since the steady state roll angle ϕ_{0s} for a constant disturbance torque L is given by

$$\frac{\phi_{0s}}{L_D}=\frac{1/20}{1\,000}=\frac{1}{-k_s K k_g L_{\delta_a} J_x} \qquad (8.2-1)$$

where

L_D = disturbance torque,

I_x = inertial moment about x axis,

$L_{\delta_a} = QSbC_{L_{\delta_a}}/I_x$,

it follows that $K_A = k_s K k_g$ must be not less than $20\,000/(344\times38.76)=1.5$. The open loop gain is now fixed at $1.5K_a = 1.5\times93.06$. If not considering compensators temporarily, then the open loop transfer function is

$$G_0=\frac{-k_s K k_g(-K_a)}{\left(\dfrac{s^2}{\omega_{ns}^2}+\dfrac{2\mu_s s}{\omega_{ns}}+1\right)s(T_a s+1)}=\frac{K_A K_a}{s\left(\dfrac{s^2}{\omega_{ns}^2}+\dfrac{2\mu_s s}{\omega_{ns}}+1\right)(T_a s+1)} \qquad (8.2-2)$$

Suppose the servo system and gyro parameters are

$$k_s=1, \quad \omega_{ns}=180\text{ rad/s}, \quad \mu_s=0.5, \quad k_g=1$$

The point at $t=16$ s is chosen to be characteristic point. Therefore, the point transfer function is

$$G_0=\frac{1.5\times93.06}{s\left(\dfrac{s^2}{180^2}+\dfrac{2\times0.5s}{180}+1\right)(0.271s+1)} \qquad (8.2-3)$$

MATLAB program 8.2-1 show a program to plot the Bode diagram. The resulting

Bode diagram is shown in Figure 8.2-2.

MATLAB program 8.2-1

```
>> num=93.06 * 1.5;
>> den=conv([1/(180*180) 2*0.5/180 1],[0.271 1 0]);
>> G0=tf(num,den);
>> margin(G0)
```

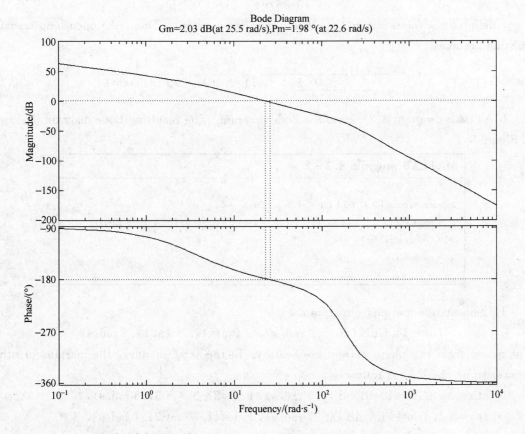

Figure 8.2-2 Bode diagram at $t=16$ s

The gain margin and phase margin are

$$Gm=2.03 \text{ dB (at } 25.5 \text{ rad/s)}, \quad Pm=1.98° \text{ (at } 22.6 \text{ rad/s)}$$

Evidently, the gain margin and phase margin are not enough. In the same manner, the other points gain margin and phase margin can be obtained as follows:

- at $t=0$ s, Gm=2.2 dB (at 34.6 rad/s), Pm=2.98° (at 30.3 rad/s),

CHAPTER 8 Autopilot Design

- at $t=8$ s, Gm=2.09 dB (at 30.7 rad/s), Pm=2.48° (at 27.1 rad/s),
- at $t=22$ s, Gm=1.93 dB (at 22.2 rad/s), Pm=1.47° (at 18.1 rad/s),
- at $t=30$ s, Gm=2.32 dB (at 180 rad/s), Pm=17.6°(at 154 rad/s),
- at $t=32.5$ s, Gm=5.13 dB (at 180 rad/s), Pm=43.2° (at 114 rad/s).

Since margins of all the above points are not enough, the compensators have to be introduced. If the introduced phase lag-lead controller is given by

$$G_C = \frac{(1/1.8)s+1}{(1/0.36)s+1} \frac{0.271s+1}{(1/40)s+1} \qquad (8.2-4)$$

then, the phase advance numerator cancels the time constant. Thus, the open loop transfer function becomes

$$G_0 = \frac{1.5 \times 93.06}{s\left(\frac{s^2}{180^2} + \frac{2 \times 0.5 s}{180} + 1\right)} \frac{(1/1.8)s+1}{(1/0.36)s+1} \frac{1}{(1/40)s+1}$$

MATLAB program 8.2-2 plots a Bode diagram. The resulting Bode diagram is shown in Figure 8.2-3.

MATLAB program 8.2-2

```
>> num=conv([0 93.06 * 1.5],[1/1.8 1]);
>> den=conv([0 1/40 1],conv([1/(180 * 180) 2 * 0.5/180 1],[1/0.36 1 0]));
>> G=tf(num,den);
>> margin(G)
```

Its gain margin and phase margin are

Gm=14.5 dB (at 75.3 rad/s),   Pm=47.7° (at 24.2 rad/s)

The gain margin and phase margin are healthy. In the same manner, the margins of other points can be obtained as follows:

- at $t=0$ s, Gm=10 dB (at 78.5 rad/s), Pm=37.5° (at 37.3 rad/s),
- at $t=8$ s, Gm=11.7 dB (at 77 rad/s), Pm=41.7° (at 31.7 rad/s),
- at $t=22$ s, Gm=18.2 dB (at 73.8 rad/s), Pm=52.4° (at 16.7 rad/s),
- at $t=30$ s, Gm=25.4 dB (at 72.3 rad/s), Pm=47.9° (at 8.17 rad/s),
- at $t=32.5$ s, Gm=28.1 dB (at 72 rad/s), Pm=42° (at 6.37 rad/s).

Thus, we use a group of control parameters for roll autopilot in stead of gain scheduling. The controller of the roll channel is

$$G = 1.5 \frac{(1/1.8)s+1}{(1/0.36)s+1} \frac{0.271s+1}{(1/40)s+1}$$

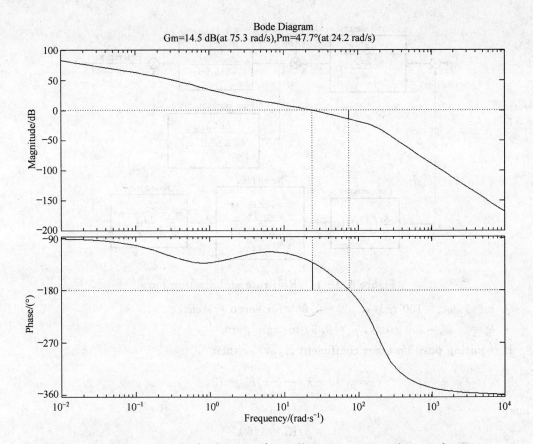

Figure 8.2-3 Bode diagram for roll position control autopilot

8.3 Autopilot Design Considering Body Flexibility

When elementarily designing autopilots, we consider the missile body to be rigid body. Actually, there is elastic deformation in missile flight. Although the frequencies of bending modes are much higher than the body natural frequency, sensors (such as rate gyros) can still sense rigid motion and elastic vibration since they have wide frequency band. These signals feedback causes bad effects on autopilots.

Figure 8.3-1 shows a roll rate autopilot design considering torsional deformation. Assume parameters as follows:

- $L_{\delta_a} = -9\,000\ \mathrm{s}^{-2}$, $L_p = -2\ \mathrm{s}^{-2}$, for rigid body,
- $K_{\mathrm{el}} = -0.000\,012\,9$, $\omega_{\mathrm{el}} = 250\ \mathrm{rad/s}$, $\mu_{\mathrm{el}} = 0.01$, for first elastic vibration mode,

CHAPTER 8 Autopilot Design

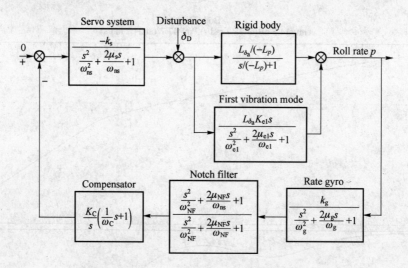

Figure 8.3 – 1 Roll rate stabilization loop

- $k_s = 1$, $\omega_{ns} = 100$ rad/s, $\mu_{ns1} = 0.65$, for servo system,
- $k_g = 1$, $\omega_g = 200$ rad/s, $\mu_g = 0.5$, for rate gyro.

If requiring position error coefficient $K_p \geqslant 75$, then

$$-k_s \frac{L_{\delta_a}}{-L_p} k_g K_C = 75$$

Thus,

$$K_C = 1/60$$

Temporarily neglect elastic mode effect. MATLAB program 8.3 – 1 gives the open loop Bode diagram. The resulting open loop Bode diagram is shown in Figure 8.3 – 2.

MATLAB program 8.3 – 1

```
>> num=9000/(2*60);
>> den=conv([0 0.5 1],conv([1/(200*200) 2*0.5/200 1],
   [1/(100*100) 2*0.65/100 1]));
>> G=tf(num,den);
>> margin(G)
```

Then the margins are

   Gm= −5.78 dB (at 76.5 rad/s),   Pm= −49.6° (at 113 rad/s)

Clearly the system is unstable.

Since the controlled plant is an inertial element, a PI (proportional plus integral) compensator is used. If choosing cross-over frequency $\omega_{CG} = 30$ rad/s, then

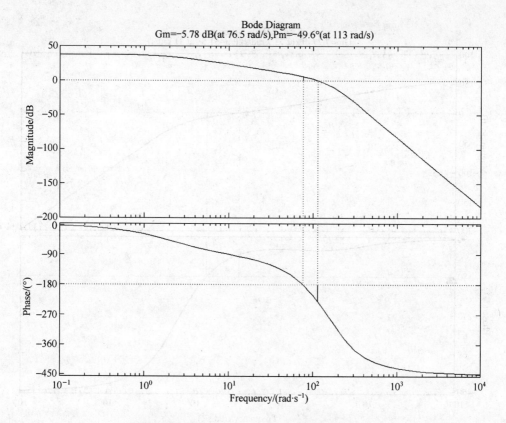

Figure 8.3-2 Open loop Bode diagram

$$75\,\frac{\dfrac{30}{\omega_C}}{30\,\dfrac{30}{2}}\approx 1$$

Therefore,

$$\omega_C = 5 \text{ rad/s}$$

MATLAB program 8.3-2 generates a Bode diagram for a roll rate autopilot with compensator. The resulting Bode diagram is shown in Figure 8.3-3.

MATLAB program 8.3-2

```
>> num=conv([0 9000/(2*60)],[0.2 1]);
>> den=conv([0.5 1 0],conv([1/(200*200) 2*0.5/200 1],
   [1/(100*100) 2*0.65/100 1]));
>> G=tf(num,den);
>> margin(G)
```

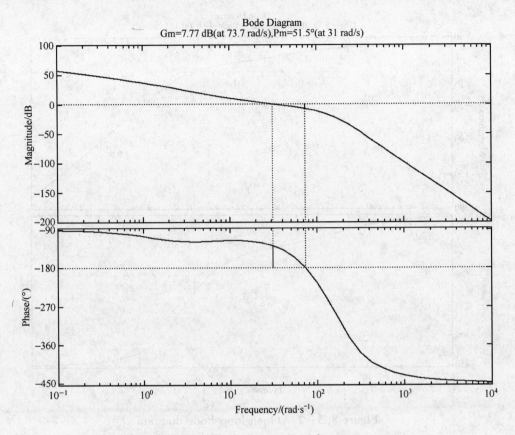

Figure 8.3-3 Bode diagram with compensator

Figure 8.3-3 presents the gain margin and phase margin as follows:

$$Gm = 7.77 \text{ dB (at 73.7 rad/s)}, \quad Pm = 51.5° \text{ (at 31 rad/s)}$$

The system has healthy margin after introducing PI compensator.

Now let us consider elastic deformation effect. MATLAB program 8.3-3 generates a Bode diagram for the system with elastic deformation. The resulting plot is shown in Figure 8.3-4.

MATLAB program 8.3-3

```
>>num=conv([(-0.0000129+1/(250*250)) (2*0.01/250-0.0000129*2) 1],conv([0 9000/(2*60)],
[0.2 1]));
>>den=conv(conv([0.5 1 0],[1/(250*250) 2*0.01/250 1]),conv([1/(200*200) 2*0.5/200 1],[1/(100*100)
2*0.65/100 1]));
>> G=tf(num,den);
>> margin(G)
```

8.3 Autopilot Design Considering Body Flexibility

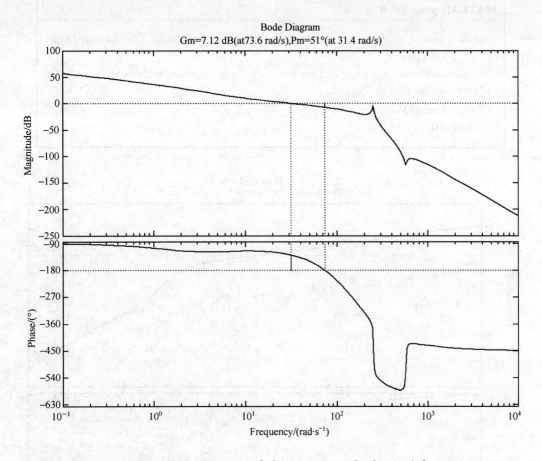

Figure 8.3 - 4 Bode diagram of the system with elastic deformation

Figure 8.3 - 4 shows the peak occurs due to small damping ratio of the first order vibration and approaches 0 dB to likely cause the system to be unstable, and furthermore a jump occurs in the phase frequency curve. In order to attenuate or cancel body structure vibration mode, a notch filter is introduced. After applying a notch filter

$$\frac{\dfrac{s^2}{250^2} + \dfrac{2\times 0.01 s}{250} + 1}{\dfrac{s^2}{250^2} + \dfrac{2\times 0.5 s}{250} + 1}$$

it cancels the poles of the first vibration. MATLAB program 8.3 - 4 gives a Bode diagram for the system with the notch filter. The resulting Bode diagram is shown in Figure 8.3 - 5.

MATLAB program 8.3 – 4

```
>>num1=conv([(-0.0000129+1/(250*250)) (2*0.01/250-0.0000129*2) 1],conv([0 9000/(2*60)],[0.2 1]));
>> den1=conv(conv([0.5 1 0],[1/(250*250) 2*0.5/250 1]),conv([1/(200*200) 2*0.5/200 1],[1/(100*100) 2*0.65/100 1]));
>> G1=tf(num1,den1);
>> margin(G1)
```

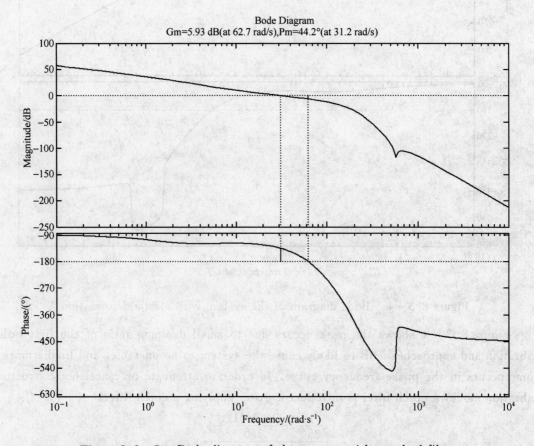

Figure 8.3 – 5 Bode diagram of the system with notched filter

Figure 8.3 – 5 shows the gain margin and phase margin as follows:

$$Gm = 5.96 \text{ dB (at 62.7 rad/s)}, \quad Pm = 44.2° \text{ (at 31.2 rad/s)}$$

The gain margin and phase margin are healthy.

8.4 Nonlinear Stability Loop Design for Roll Channel

Electric actuators are used in missiles. There are two manners in electric actuators. They are direct manner and indirect manner. For indirect manner, the motor of the actuator is running at constant speed, and the output of the actuator is controlled by the clutches on-off instead of directly controlling the voltage of the motor. The clutch is a typical nonlinear element and represents a relay characteristic. Figure 8.4 – 1 shows a block diagram of roll position stabilization loop with a nonlinear element.

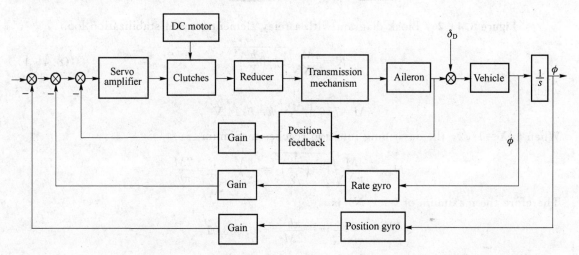

Figure 8.4 – 1 Block diagram of roll position stabilization loop with a nonlinear element

The servo amplifier synthesizes position gyro, rate gyro and aileron feedback signals. According to its polarity, the synthesized signal of the amplifier output controls the clutch 1 or 2 on-off. Before the stabilization loop begins to work, the DC permanent magnet motor runs. We do not directly control DC motor armature voltage. The motor runs at constant speed. By the clutches, the motor are connected with the aileron. After a clutch is on, through the reducer and transmission mechanism, the motor drives the aileron to deflect. The forward path may be equivalent to a relay and delay elements as shown in Figure 8.4 – 2.

1. Fin loop design

The describing function of the relay element is

CHAPTER 8 Autopilot Design

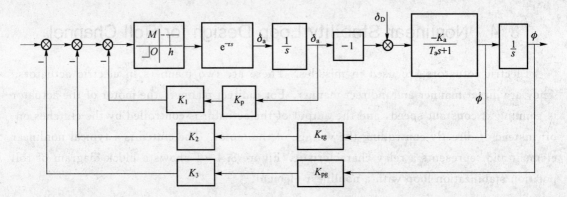

Figure 8.4 - 2 Block diagram with a relay element for roll stabilization loop

$$N = \frac{4M}{\pi X}\sqrt{1-\left(\frac{h}{X}\right)^2} \tag{8.4-1}$$

where

$$M = 73(°)/s, \quad h = 0.14 \text{ V}$$

When $h/X = 1/\sqrt{2}$, the describing function reaches maximum

$$N_{max} = \frac{4M}{\pi X}\sqrt{1-\left(\frac{1}{\sqrt{2}}\right)^2} = \frac{4M}{\pi\sqrt{2}h}\frac{1}{\sqrt{2}} = \frac{2M}{\pi h}$$

Therefore the maximum of $(-1/N)$ is

$$-\frac{1}{N_{max}} = -\frac{\pi h}{2M} = -0.00301$$

The remaining element transfer function of the fin loop is

$$G_F = \frac{K_1 K_p e^{-\tau s}}{s} \tag{8.4-2}$$

where $\tau = 13 \times 10^{-3}$ s, $K_p = 1$ V/(°).

Therefore, the magnitude-phase characteristic of the transfer function (8.4-2) is given by

$$G_F(j\omega) = \frac{K_1 K_p e^{-j\tau\omega}}{j\omega} \tag{8.4-3}$$

It has infinite cross-over points in the negative real axis. But maximum magnitude corresponds to $-\pi$, i.e.

$$-(\omega_C \tau + \pi/2) = -\pi \tag{8.4-4}$$

Then

$$\omega_C = \pi/(2\tau) \tag{8.4-5}$$

Thus,

8.4 Nonlinear Stability Loop Design for Roll Channel

$$|G_F(j\omega_C)| = K_1 K_p/\omega_C \qquad (8.4-6)$$

In order to avoid self-oscillation at ω_C, the system requires

$$K_1 K_p/\omega_C < 0.003\,01 \qquad (8.4-7)$$

If choose $K_1 = 0.23$, then

$$\frac{K_1 K_p}{\omega_C} = \frac{0.23 \times 1}{\frac{3.14}{2 \times 13 \times 10^{-3}}} = 0.001\,9 < 0.003\,01$$

Its margin is $0.003\,01/0.001\,9 = 1.56$.

Thus, MATLAB program 8.4-1 generates $(-1/N)$ and $G_F(j\omega)$ curves. The resulting plot is shown in Figure 8.4-3.

```
MATLAB program 8.4-1
>> M=73;h=0.14;
>> X=0.145;
>> for i=1:40
        X=X+0.05;
        x(i)=-1/(4*M/(pi*X)*sqrt(1-(h/X)^2));
        y(i)=0;
    end
>> plot(x,y,'r-');
>> hold on
>> n1=[0.23];d1=[1 0]; G1=tf(n1,d1);
>> tau=0.013;[n2,d2]=pade(tau,4);G2=tf(n2,d2);
>> G3=G1*G2;
>> nyquist(G3,{10,1000});
>> axis([-0.01 0.01 -0.01 0.01]);
```

2. Stabilization loop design

In stabilization loop design, first let the fin loop be approximately $1/(K_1 K_p)$. Then Figure 8.4-2 reduces to Figure 8.4-4.

From Figure 8.4-2, it is easy to obtain the following closed loop transfer function:

$$\frac{\phi}{\delta_D} = \frac{\dfrac{K_1 K_p}{K_3 K_{pg}}}{\dfrac{T_a K_1 K_p}{K_a K_3 K_{pg}} s^2 + \dfrac{K_1 K_p + K_a K_2 K_{rg}}{K_a K_3 K_{pg}} s + 1} \qquad (8.4-8)$$

Here, assume

CHAPTER 8 Autopilot Design

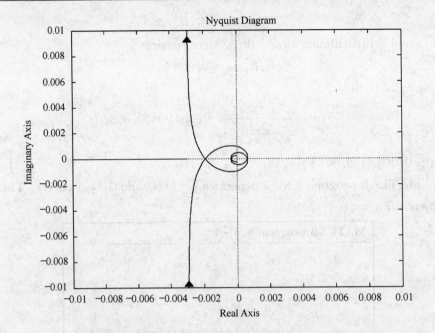

Figure 8.4-3 $(-1/N)$ and $G_F(j\omega)$ curves

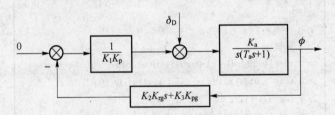

Figure 8.4-4 Simplified block diagram

$K_a = 361.93 \text{ s}^{-1}$, $T_a = 0.025\ 7 \text{ s}$, $K_{rg} = 1 \text{ V}/[(°) \cdot \text{s}]$, $K_{pg} = 1 \text{ V}/(°)$

If the maximum disturbance moment corresponds to equivalent deflection $\delta_D = 3°$, and the permissible maximum roll angle is $2°$, then

$$\frac{2}{3} = \frac{K_1 K_p}{K_3 K_{pg}} = \frac{0.23 \times 1}{K_3 \times 1} \tag{8.4-9}$$

Thus,

$$K_3 = 0.345$$

The natural frequency of equation (8.4-8) is

$$\omega_n = \left(\frac{T_a K_1 K_p}{K_a K_3 K_{pg}}\right)^{-\frac{1}{2}} = 145.34$$

If require that the stabilization damping ratio is 0.5, then

$$\frac{K_1 K_p + K_a K_2 K_{rg}}{K_a K_3 K_{pg}} = 2 \times 0.5/145.34$$

Therefore,

$$K_2 = 0.001\,738$$

In order to verify self-oscillation, we incorporate feedback loops. Figure 8.4-2 becomes Figure 8.4-5.

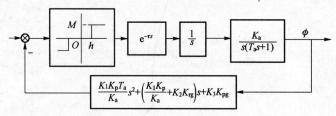

Therefore we obtain resultant transfer function of linear elements and the delay element below.

$$G = \left[\frac{K_1 K_p T_a}{K_a} s^2 + \left(\frac{K_1 K_p}{K_a} + K_2 K_{rg} \right) s + K_3 K_{pg} \right] \frac{K_a}{s^2 (T_a s + 1)} e^{-\tau s} =$$

$$\left[\frac{K_1 K_p T_a}{K_a K_3 K_{pg}} s^2 + \left(\frac{K_1 K_p + K_a K_2 K_{rg}}{K_a K_3 K_{pg}} \right) s + 1 \right] \frac{K_3 K_{pg} K_a}{s^2 (T_a s + 1)} e^{-\tau s}$$

Through MATLAB program 8.4-2, $(-1/N)$ and $G(j\omega)$ curves are plotted. Figure 8.4-6 presents the resulting plot.

MATLAB program 8.4-2

```
>> M=73;h=0.14;
>> X=0.145;
>> for i=1:40
       X=X+0.05;
       x(i)=-1/(4*M/(pi*X)*sqrt(1-(h/X)^2));
       y(i)=0;
   end
>> plot(x,y,'r-');
>> hold on
>> n1=conv([1/145.34^2 2*0.5/145.34 1],[0 0 0.345*361.93]);
>> d1=[0.0257 1 0 0];G1=tf(n1,d1);
>> tau=0.013;[n2,d2]=pade(tau,4);G2=tf(n2,d2);
>> G3=G1*G2;
>> nyquist(G3,{10,1000});
>> axis([-0.01 0.01 -0.01 0.01]);
```

Figure 8.4-6 shows that there are no intersecting points between $(-1/N)$ and $G(j\omega)$ curves. Self-oscillation does not occur.

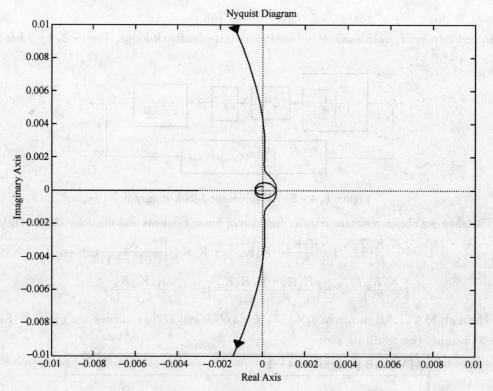

Figure 8.4-6 $(-1/N)$ and $G(j\omega)$ curves

8.5 Acceleration Control System Design

Now let us consider a missile with rear controls whose yaw derivatives at $u_0 = 500$ m/s are

$$Y_v = -3, \quad N_v = 1, \quad Y_{\delta_r} = 180, \quad N_{\delta_r} = -500, \quad N_r = -3$$

From equations (5.6-14) and (5.6-16),

$$\frac{\Delta r(s)}{\Delta \delta_r(s)} = \frac{N_{\delta_r} s - N_{\delta_r} Y_v + N_v Y_{\delta_r}}{s^2 - (Y_v + N_r)s + Y_v N_r + u_0 N_v} = \frac{-500(s+2.64)}{s^2 + 6s + 509} \quad (8.5-1)$$

$$\frac{a_y(s)}{\Delta \delta_r(s)} = \frac{Y_{\delta_r} s^2 - Y_{\delta_r} N_r s - u_0 (N_{\delta_r} Y_v - N_v Y_{\delta_r})}{s^2 - (Y_v + N_r)s + Y_v N_r + u_0 N_v} = \\ \frac{180s^2 + 540s - 660\,000}{s^2 + 6s + 509} \quad (8.5-2)$$

8.5 Acceleration Control System Design

Equation (8.5-2) divided by equation (8.5-1) yields

$$\frac{a_y(s)}{\Delta r(s)} = \frac{-0.36s^2 - 1.08s + 1320}{s + 2.64} \quad (8.5-3)$$

The rate gyro feedback is used to increase damping ratio. Assume the accelerometer is put 0.5 m ahead of the center of gravity. The pure integrator is introduced to eliminate steady error. Assume the servo system transfer function is

$$\frac{-0.007}{(1/180^2)s^2 + 2 \times 0.5/180s + 1} \quad (8.5-4)$$

Figure 8.5-1 shows control configuration.

Referring to Figure 8.5-1, the open transfer function of the rate loop is written as

$$G_{or} = K_{rg} \frac{0.007}{(1/180^2)s^2 + 2 \times 0.5/180s + 1} \frac{500(s + 2.64)}{s^2 + 6s + 509} \quad (8.5-5)$$

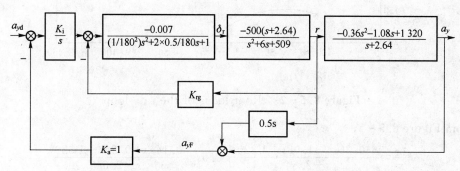

Figure 8.5-1 Lateral autopilot

MATLAB program 8.5-1 plots a root locus. The resulting root locus is shown in Figure 8.5-2.

MATLAB program 8.5-1
```
>> n1=[0.007];d1=[1/(180*180) 1/180 1];G1=tf(n1,d1);
>> n2=[500 1320];d2=[1 6 509];G2=tf(n2,d2);
>> G3=G1*G2;
>> rlocus(G3); axis([-150,150,-200,200]);
```

After simulation computation, it is found that when $K_{rg} = 8$ the damping ratio is 0.76. The following MATLAB commands gives the transfer function of the closed loop for rate channel.

```
>> Gcr=feedback(G3,8); zpk(Gcr);
Zero/pole/gain:
               113400 (s+2.64)
      -------------------------------------
      (s^2+39.75s +687.1) (s^2+146.2s +2.749e004)
```

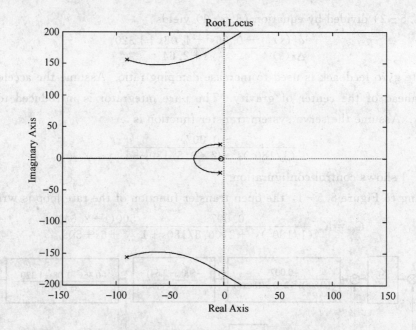

Figure 8.5-2 Root locus of the rate loop

From Figure 8.5-1,

$$\frac{a_{rF}}{r} = \frac{-0.36s^2 - 1.08s + 1\,320}{s + 2.64} + 0.5s = \frac{0.14s^2 + 0.24s + 1\,320}{s + 2.64}$$

Then Figure 8.5-1 becomes Figure 8.5-3.

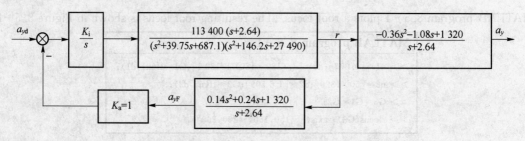

Figure 8.5-3 Simplified block diagram

Thus, the open loop transfer function of the acceleration loop is

$$G_{oa} = \frac{K_i}{s} \frac{113\,400(0.14s^2 + 0.24s + 1\,320)}{(s^2 + 39.75s + 687.1)(s^2 + 146.2s + 27\,490)}$$

MATLAB program 8.5-2 generates a root locus for the above transfer function. The resulting plot is shown in Figure 8.5-4.

8.5 Acceleration Control System Design

MATLAB program 8.5 - 2

```
>> n2=[1];d2=[1 0];G2=tf(n2,d2);
>> n3=[113400*0.14 113400*0.24 113400*1320];
>> d3=conv([1 39.75 687.1],[1 146.2 27490]);G3=tf(n3,d3);
>> G4=G2*G3;rlocus(G4);
>> axis([-100 100 -300 300]);
```

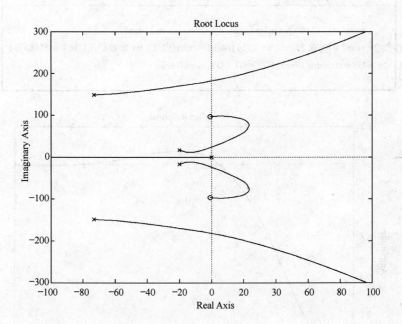

Figure 8.5 - 4 Root locus of the acceleration loop

By trial-and-error, $K_i = 1$ is appropriate. MATLAB program 8.5 - 3 gives the transfer function of r/a_{yd}.

MATLAB program 8.5 - 3

```
>> n1=[1];d1=[1 0];G1=tf(n1,d1);
>> n2=[113400 113400*2.64];
>> d2=conv([1 39.75 687.1],[1 146.2 27490]);G2=tf(n2,d2);
>> G3=G1*G2;
>> n4=[0.14 0.24 1320];d4=[1 2.64];G4=tf(n4,d4);
>> G5=feedback(G3,G4);zpk(G5);
   Zero/pole/gain:
                      113400 (s+2.64)^2
   ----------------------------------------------------------
   (s+21.4) (s+2.64) (s^2+18.94s+254.8) (s^2+145.6s+2.745e004)
```

Thus,

$$\frac{a_y}{a_{yd}} = \frac{113\,400(-0.36s^2 - 1.08s + 1\,320)}{(s+21.4)(s^2+18.94s+254.8)(s^2+145.6s+27\,450)} \quad (8.5-6)$$

MATLAB program 8.5 - 4 generates a step response of the acceleration. The resulting step response is shown in Figure 8.5 - 5.

MATLAB program 8.5 - 4

```
>> num1=[113400];den1=[1 21.4];G1=tf(num1,den1);
>> num2=[-0.36 -1.08 1320];den2=conv([1 18.94 254.8],[1 145.6 27450]);
>> G2=tf(num2,den2);G3=G1*G2;step(G3);
```

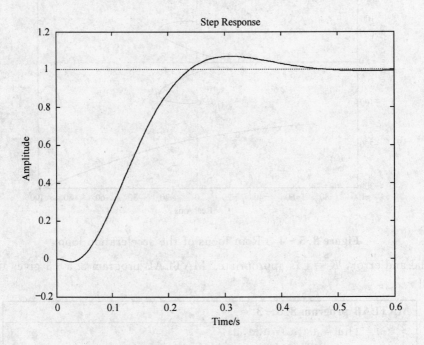

Figure 8.5 - 5 Acceleration step response

If the change of aerodynamic parameters is not too great, it is possible for the fixed gains to suit the change. If there is great change in stability derivatives, the gain change or called gain scheduling is required. The control gains vary with the dynamic pressure and /or Mach number.

8.6 Longitudinal Control System Design for Cruise Missile

For a shore to ship missile, flight altitude is 100 m, and flight velocity is 312.7 m/s. In this case, the state equation is given by

$$\begin{bmatrix} \Delta \dot{u} \\ \Delta \dot{q} \\ \Delta \dot{\alpha} \\ \Delta \dot{\theta} \end{bmatrix} =
\begin{bmatrix} -0.006\,01 & 0 & -0.064\,7+9.8 & -9.8 \\ -0.007\,697+0.571\,6\times 0.000\,071\,07 & -2.125-0.571\,6 & 0.571\,6\times 0.943-45.397 & 0 \\ -0.000\,071\,07 & 1 & -0.943 & 0 \\ 0 & 1 & 0 & 0 \end{bmatrix}
\begin{bmatrix} \Delta u \\ \Delta q \\ \Delta \alpha \\ \Delta \theta \end{bmatrix} +
\begin{bmatrix} 0 \\ -68.13 \\ -0.113\,5 \\ 0 \end{bmatrix} \delta_e \quad (8.6-1)$$

Example 4.5-3 presents that the system has a pole, 0.026, in right-half-plane. It is unstable. The pitch angle feedback may stabilize the long-period mode. The rate feedback may increase the damping ratio. The pitch-attitude hold will be used as the inner loop of altitude hold. The control configuration is shown in Figure 8.6-1.

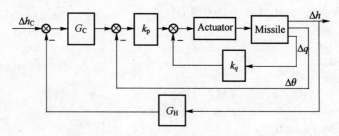

Figure 8.6-1 Altitude-hold autopilot

For convenience, the equation (3.4-20) is rewritten as

$$\Delta \dot{z}_E = -\Delta u \sin\theta_0 + \Delta w \cos\theta_0 - u_0 \Delta\theta \cos\theta_0 \quad (8.6-2)$$

Let $\theta_0 = 0$. Then the equation becomes

CHAPTER 8 Autopilot Design

$$\Delta \dot{z}_E = \Delta w - u_0 \Delta \theta = u_0 \Delta \alpha - u_0 \Delta \theta \qquad (8.6-3)$$

Thus,

$$\Delta \dot{h} = -\Delta \dot{z}_E = u_0 \Delta \theta - u_0 \Delta \alpha \qquad (8.6-4)$$

By augmenting state equation, the transfer function $\Delta h/\Delta \theta$ can be obtained. MATLAB program 8.6-1 generates the transfer function.

MATLAB program 8.6-1

```
>> a1=[-0.00601 0 -0.0647+9.8 -9.8 0
-0.007697+0.5716*0.00007107 -2.125-0.5716 0.5716*0.943-45.397 -0.5716*0 0
-0.00007107 1 -0.943 0 0
0 1 0 0 0
0 0 -312.7 312.7 0];
>> b1=[0;-68.13;-0.1135;0;0];
>> c1=[0 1 0 0 0;0 0 0 1 0;0 0 0 0 1];d1=[0;0;0];
>> [zxita,pxita,kxita]=ss2zp(a1,b1,c1(2,:),d1(2,:));
   zxita =
      -0.0067
      -0.8676
            0
   pxita =
            0
      -1.8198+6.6401i
      -1.8198-6.6401i
       0.0260
      -0.0321
   kxita =
      -68.1300
[zh,ph,kh]=ss2zp(a1,b1,c1(3,:),d1(3,:))
   zh =
      -24.2177
       21.5211
       -0.0060
   ph =
            0
      -1.8198 +6.6401i
      -1.8198 - 6.6401i
       0.0260
      -0.0321
   kh =
       35.4915
```

8.6 Longitudinal Control System Design for Cruise Missile

Therefore,

$$\frac{\Delta\theta}{\Delta\delta_e} = \frac{-68.13(s+0.006\,7)(s+0.867\,6)}{(s+1.819\,8+6.640\,1i)(+1.819\,8-6.640\,1i)(s+0.032\,1)(s-0.026)} \quad (8.6-5)$$

$$\frac{\Delta h}{\Delta\delta_e} = \frac{35.491\,5(s+24.217\,7)(s+0.006)(s-21.521\,1)}{s(s+1.819\,8+6.640\,1i)(+1.819\,8-6.640\,1i)(s+0.032\,1)(s-0.026)} \quad (8.6-6)$$

The equation (8.6-6) divided by equation (8.6-5) is

$$\frac{\Delta h}{\Delta\theta} = \frac{-0.520\,9(s+24.217\,7)(s+0.006)(s-21.521\,1)}{s(s+0.006\,7)(s+0.867\,6)} \quad (8.6-7)$$

The plant matrices will be renamed **a**, **b** and so on, and augmented with a simple-lag elevator model of time-constant $1/75$ s. In view of negative feedback, a negative sign is added to the servo. The transfer function of the servo is $-75/(s+75)$. MATLAB program 8.6-2 extracts state-space matrices of the system.

From trial and error, it is found that $k_q = 0.2$, $k_p = 1$ will result in a reasonable autopilot. MATLAB program 8.6-3 generates the zero-pole model of the attitude control closed system.

MATLAB program 8.6-2

```
>> a=[-0.006010 -0.0647+9.8 -9.8;
-0.007697+0.5716*0.00007107 -2.125-0.5716 0.5716*0.943-0.5716*0 -45.397 -0.5716*0;
-0.00007107 1 -0.943+0 0;
0 1 0 0];
b=[0;-68.13;-0.1135;0];
c=[0 1 0 0;0 0 0 1];
d=[0;0];
plant=ss(a,b,c,d);
>> aa=[-75];ba=[75];ca=[-1];da=[0];
>> actua=ss(aa,ba,ca,da);
>> sys1=series(actua,plant);
>> [aa,bb,cc,dd]=ssdata(sys1);
```

CHAPTER 8 Autopilot Design

> **MATLAB program 8.6 – 3**
>
> ```
> >> ac1=aa-bb*[0. 2 1]*cc;
> >> [z,p,k]=ss2zp(ac1,bb,cc(2,:),dd(2,:),1)
> z =
>     -0.0067
>     -0.8676
> p =
>     -58.4472
>     -9.8470 +7.4401i
>     -9.8470 - 7.4401i
>     -0.0060
>     -0.4984
> k =
>     5.1098e+003
> ```

Then the closed loop transfer function with the feedbacks $k_q = 0.2$, $k_p = 1$ is written as

$$\frac{\Delta\theta}{\Delta\theta_C} = \frac{5\,109.8(s+0.006\,7)(s+0.867\,6)}{(s+58.447\,2)(s+9.847+7.440\,1i)(s+9.847-7.440\,1i)(s+0.006)(s+0.498\,4)}$$

(8.6 – 8)

The equation (8.6 – 7) divided by equation (8.6 – 8) is

$$\frac{\Delta h}{\Delta\theta_C} = \frac{-2\,661.7(s+24.217\,7)(s-21.521\,1)}{s(s+58.447\,2)(s+9.847+7.440\,1i)(s+9.847-7.440\,1i)(s+0.498\,4)}$$

(8.6 – 9)

Use RLTOOL of MATLAB (SISO Design Tool) to design altitude hold autopilot. MATLAB program 8.6 – 4 gives the SISO Design Tool. The SISO Design Tool is shown in Figure 8.6 – 2.

> **MATLAB program 8.6 – 4**
>
> ```
> >> nh=conv([-2661.7 -2661.7*24.2177],[1 -21.5211]);
> >> dh=conv(conv([1 9.847+7.4401i],[1 9.847-7.4401i]),conv([1 58.4472 0],[0 1 0.4984]));
> >> sysh=tf(nh,dh);
> >> rltool(sysh);
> ```

The result of simulation test indicates that adding a compensator $0.020\,8(1+s)$ will yield a satisfactory step response. The step response is shown in Figure 8.6 – 3.

8.6 Longitudinal Control System Design for Cruise Missile

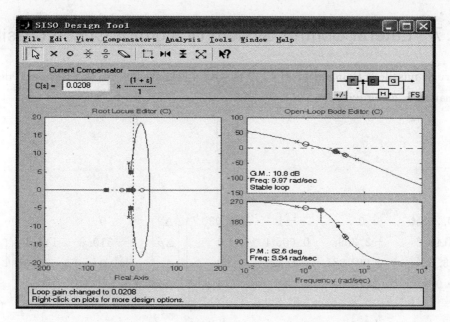

Figure 8.6 − 2 SISO Design Tool of altitude hold

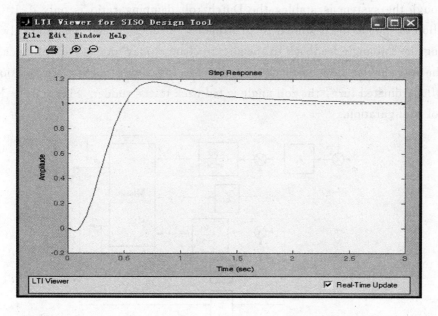

Figure 8.6 − 3 Step-response of the altitude holding autopilot

CHAPTER 8 Autopilot Design

8.7 Lateral Control System Design for Cruise Missile

A cruise missile is flying at height 100 m with a velocity 300 m/s. The lateral-directional state equation is given by

$$\begin{bmatrix} \Delta \dot{\beta} \\ \Delta \dot{p} \\ \Delta \dot{r} \\ \Delta \dot{\phi} \end{bmatrix} = \begin{bmatrix} -0.2984 & 0 & -1 & -0.0307 \\ -148.6 & -2.309 & 0.6054 & 0 \\ 53.62 & -0.001981 & -1.337 & 0 \\ 0 & 1 & 0 & 0 \end{bmatrix} \begin{bmatrix} \Delta \beta \\ \Delta p \\ \Delta r \\ \Delta \phi \end{bmatrix} + \begin{bmatrix} 0 & 0 \\ -310.8 & 118.1 \\ 0 & -32.66 \\ 0 & 0 \end{bmatrix} \begin{bmatrix} \Delta \delta_a \\ \Delta \delta_r \end{bmatrix}$$

(8.7-1)

It is easy to obtain the characteristic roots as follows:
$$-2.3573; \quad -0.7735 \pm 7.3072i; \quad -0.00401$$

Although the system is stable, the Dutch roll damping ratio is only 0.1. Since the configuration of the missile is similar to an airplane, the ailerons and rudder are always used. Introducing the roll angle feedback to the aileron channel may increase spiral mode stability. Adding the roll angular rate feedback may increase roll flying quality. In addition, in the interest of coordinated turn, the roll angle is fed back to the rudder. Figure 8.7-1 indicates the control configuration.

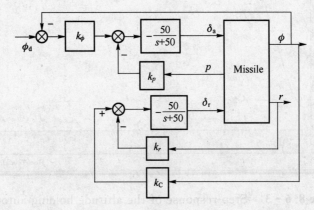

Figure 8.7-1 Block diagram of inner loop of lateral control

8.7 Lateral Control System Design for Cruise Missile

The plant matrices will be renamed **a**, **b** and so on, and augmented with a simple-lag rudder model of time-constant 1/50 s. In view of negative feedback, we add a negative sign to the servo. The transfer function of the servo is $-50/(s+50)$. MATLAB program 8.7-1 extracts state-space matrices of the system.

MATLAB program 8.7-1

```
>> a=[-0.2984 0 -1 0.0307
     -148.6 -2.309 0.6054 0
     53.62 -0.001981 -1.337 0
     0 1 0 0];
>> b=[0 0;-310.8 118.1;0 -32.66;0 0];
>> c=[1 0 0 0;0 1 0 0;0 0 1 0;0 0 0 1];
>> d=[0 0;0 0;0 0;0 0];
>> plant=ss(a,b,c,d);
>> aa=[-50 0;0 -50];ba=[50 0;0 50];
>> ca=[-1 0;0 -1];da=[0 0;0 0];
>> actua=ss(aa,ba,ca,da);
>> sys1=series(actua,plant);
>> [a1,b1,c1,d1]=ssdata(sys1);
```

From trial and error, it is found that $k_\phi = 1.5$, $k_p = 0.1$, $k_r = 0.3$, $k_C = 0.0105$ will result in a reasonable autopilot. MATLAB program 8.7-2 generates step responses. Figure 8.7-2 shows the step response for the roll angle ϕ. Figure 8.7-3 shows the step response for the sideslip angle β.

MATLAB program 8.7-2

```
>> ac1=a1-b1*[0 0.1 0 1.5;0 0 0.3 -0.0105]*c1;
>> step(ac1,b1(:,1)*1.5,c1(4,:),d1(4,:),1);
>> step(ac1,b1(:,1)*1.5,c1(1,:),d1(1,:),1);
```

Figure 8.7-3 shows that it takes 0.7 s to let the sideslip angle β become zero. Therefore, the target of coordinated turn has been achieved.

MATLAB program 8.7-3 gives of the Bode diagram of open loop of roll angle. Figure 8.7-4 indicates the Bode diagram.

CHAPTER 8 Autopilot Design

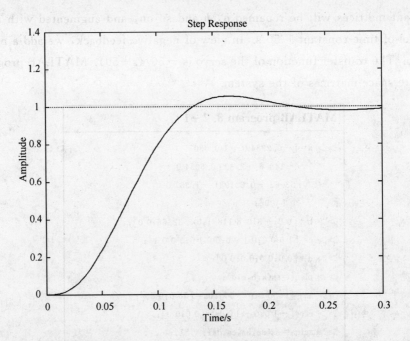

Figure 8.7 – 2 Step response for roll angle ϕ

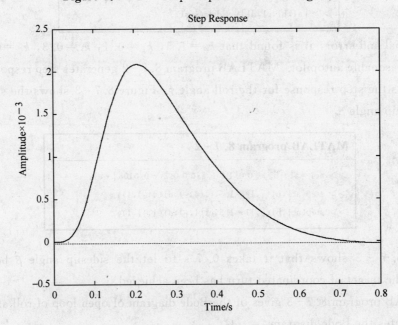

Figure 8.7 – 3 Step response for sideslip angle β

8.7 Lateral Control System Design for Cruise Missile

MATLAB program 8.7 - 3

```
>> ac2=a1-b1*[0 0.1 0 0;0 0 0.3 -0.0105]*c1;
>> [n1,d1]=ss2tf(ac1,b1(:,1)*1.5,c1(4,:),0,1);
>> margin(n1,d1);
```

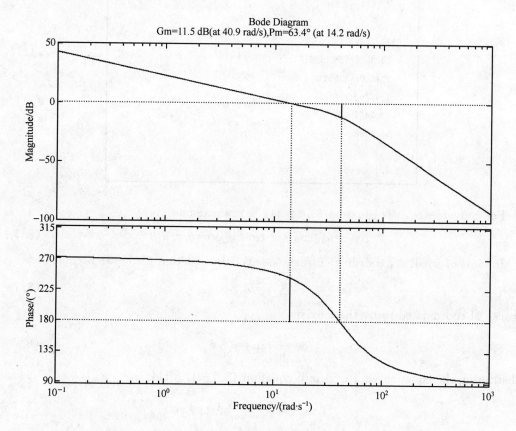

Figure 8.7 - 4 Bode diagram of open loop of roll angle

Figure 8.7 - 4 shows Gm=11.5 dB, Pm=63.4°.

MATLAB program 8.7 - 4 generates the closed loop transfer function for ϕ/ϕ_d. That is

$$\frac{\phi}{\phi_d} = \frac{233\,10(s+36.700\,8)(s+7.467\,3\pm4.670\,8i)}{(s+14.354\,9\pm28.038\,3i)(s+7.432\pm4.658\,1i)(s+23.745\,1)(s+36.625\,6)}$$

(8.7 - 2)

CHAPTER 8 Autopilot Design

MATLAB program 8.7 - 4

```
>> [z,p,k]=ss2zp(ac1,b1(:,1)*1.5,c1(4,:),0,1)
z =
   -36.7008
   -7.4673+4.6708i
   -7.4673-4.6708i
p =
   -14.3549+28.0383i
   -14.3549-28.0383i
   -7.4320+4.6581i
   -7.4320-4.6581i
   -23.7451
   -36.6256
k =
   2.3310e+004
```

For convenience, the equation (3.4 - 24) is rewritten as

$$\Delta \dot{y}_E = u_0 \psi \cos\theta_0 + \Delta v = u_0 \psi \cos\theta_0 + u_0 \beta \tag{8.7-3}$$

In view of small θ_0, coordinated turn ($\beta \approx 0$), the equation (8.7 - 3) becomes

$$\Delta \dot{y}_E = u_0 \psi \tag{8.7-4}$$

Because of coordinated turn, there exists

$$\dot{\psi} = \frac{g}{u_0} \tan\phi \approx \frac{g}{u_0} \phi \tag{8.7-5}$$

Substituting the equation (8.7 - 5) into equation (8.7 - 4) yields

$$\Delta \ddot{y}_E = u_0 \dot{\psi} = u_0 \frac{g}{u_0} \phi = g\phi \tag{8.7-6}$$

Taking Laplace transformation of the equation (8.7 - 6) yields

$$\frac{\Delta y_E}{\phi} = \frac{g}{s^2} \tag{8.7-7}$$

Thus, the transfer function $\Delta y_E / \phi_d$ is

$$\frac{\Delta y_E}{\phi_d} = \frac{9.8 \times 233\,10(s+36.700\,8)(s+7.467\,3 \pm 4.670\,8i)}{s^2(s+14.354\,9 \pm 28.038\,3)(s+7.432 \pm 4.658\,1i)(s+23.745\,1)(s+36.625\,6)} \tag{8.7-8}$$

About lateral-direction deviation control, Figure 8.7 - 5 shows a control configuration.

8.7 Lateral Control System Design for Cruise Missile

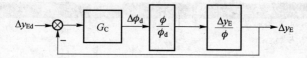

Figure 8.7 – 5 Block diagram of lateral-direction deviation control

Use RLTOOL of MATLAB (SISO Design Tool) to design lateral-direction deviation control autopilot. MATLAB program 8.7 – 5 gives the SISO Design Tool. The SISO Design Tool is shown in Figure 8.7 – 6.

MATLAB program 8.7 – 5
```
>> n=conv(conv([1 7.4673+4.6708i],[1 7.4673-4.6708i]),[0 9.8*23310 9.8*23310*36.7008]);
>> d1=conv(conv([1 14.3549+28.0383i],[1 14.3549-28.0383i]),conv([1 7.432+4.6581i],[1 7.432-4.6581i]));
>> d2=conv([1 23.7451 0 0],[0 0 1 36.6256]);
>> d=conv([d1],[d2]);sys=tf(n,d);
>> rltool(sys);
```

The result of simulation test indicates that adding a compensator $(1+s)$ will yield a satisfactory step response. Figure 8.7 – 7 shows the step response.

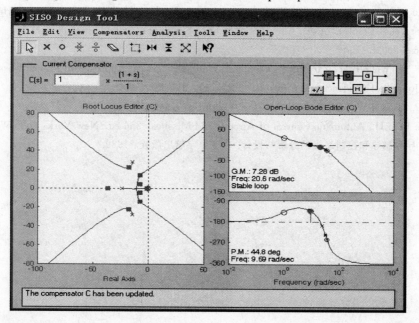

Figure 8.7 – 6 SISO Design Tool of lateral-direction deviation control

CHAPTER 8 Autopilot Design

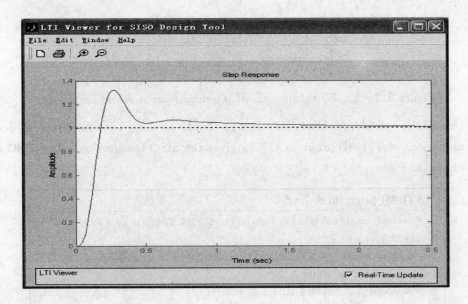

Figure 8.7-7 Step-response of the lateral-direction deviation autopilot

References

[1] Robert C Nelson. Flight Stability and Automatic Control. New York: McGraw-Hill, 1989.
[2] Katsuhiko Ogata. Modern Control Engineering. 4th ed. Englewood Cliffs, N. J. : Prentice Hall, 2002.
[3] Brian L Stevens, Frank L Lewis. Aircraft Control and Simulation. 2nd ed. Hoboken, N. J. : John Wiley & Sons, Inc. , 2003.
[4] Nesline F W, Wells B H, Zarchan P. Combined Optimal/Classical Approach to Robust Missile Autopilot Design. Journal of Guidance and Control, 1982, 4(3): 316-332.
[5] Garnell P. Guided Weapon Control Systems. Oxford: Pergamon Press, 1980.
[6] Blakelock John H. Automatic Control of Aircraft and Missiles. 2nd ed. New York: John Wiley, 1991.
[7] 陈佳实. 导弹制导和控制系统的分析与设计. 北京: 宇航出版社, 1989.

CHAPTER 9
Command Guidance Systems

9.1 Principle of Command Guidance

9.1.1 Introduction

The remote control guidance is widely used by anti-airplane missiles (such as Nike, SA—2, Rapier) and anti-tank missiles (such as Cobra, Hot, Malkara). At present the remote control guidance is classified into two classes: remote control command guidance and beam-riding guidance.

In command guidance, the off-board guidance station, which is in a truck-carrier on the ground, a ship-carrier on the sea, or an airplane-carrier in the air, provides guiding intelligence to a missile; and then proper manipulation yields guiding commands to steer the missile at a target till the missile destroy the target.

In beam-riding guidance, the guidance station transmits radio beam or laser beam at a target instead of guidance commands; the missile always flies on the wave beam and sustains in the center line of the wave beam through sensing the error off the center line till hitting the target. As the guidance method is simple, it is widely used in early missiles. But there are deficiencies as follows:

① Since the wave beam must aim at the target forever in guiding process, it is not only easy to stick one's chin out, but also restrict its own maneuverability;

② In order to reduce guidance errors, the wave beam is required to be narrower. This increases the difficulty of capturing target; in addition, the missile is easy to move out of the wave beam.

CHAPTER 9 Command Guidance Systems

Therefore, the guidance method is rarely adopted at present.

The radio command guidance is the remote control command guidance widely used. The guidance commands produced by the guidance station outside missile will be transmitted to missile through the radio wave. Two radars are used to track missile and target respectively, or a radar is used to track both target and missile.

The guidance radar of guidance station can measure the position information of target and missile. The command generator in guidance station chooses a proper guidance law to produce guidance commands. The commands transmitted by the command transmitter are received by the receiver of an on-board transponder. The received commands are decoded to yield control signals, which will be fed to autopilot of the missile. The autopilot controls flight trajectory of the missile. Figure 9.1 – 1 shows control configuration of a command guidance system.

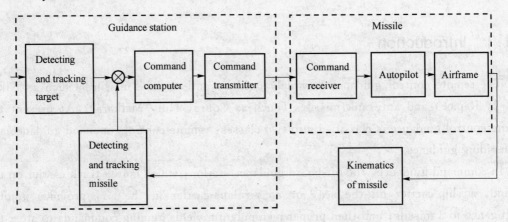

Figure 9.1 – 1 Block diagram of a command guidance system

Evidently, the command guidance system consists of the off-board guidance station and the on-board receiving and controlling devices in missile. The guidance station is responsible of tracking and measuring target and missile, producing guidance commands and transmitting guidance intelligence. Indeed, a single radar may detect target and missile simultaneously. But how to distinguish target and missile? There is a transponder on board. After a missile is launched, the guidance station will transmit interrogating signals. The on-board transponder returns responding signals, while the target can not answer the interrogating signals. Therefore, measuring the transmission time from interrogation to response may compute the distance from the station to the missile.

A typical radio command guidance system is shown in Figure 9.1 – 2.

9.1 Principle of Command Guidance

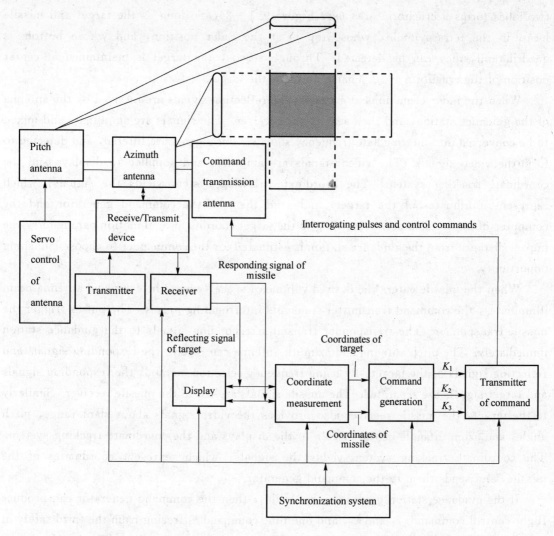

Figure 9.1 – 2 Single radar radio command guidance system

Synchronization system provides a reference pulse to subsystems of the radar for coordination. The pitch antenna and the azimuth antenna produce tabular wave beams respectively. Triggered by synchronization pulse, the beams scan in space. The pitch lobe, which is generated by the pitch antenna, is narrower (for instance 2°) in vertical direction, but wider (for instance 10°) in horizontal direction. The pitch lobe scans through a certain angle (for instance 20°) in vertical direction. The azimuth lobe is narrower (for instance 2°) in horizontal direction, but wider (for instance 10°) in vertical direction. The azimuth lobe scans through a certain angle (for instance 20°) in horizontal direction. The scanning of the

CHAPTER 9 Command Guidance Systems

two lobes forms a cruciform area (see Figure 9.1 - 2). So long as the target and missile locate in the tetra-pyramid, whose top is at the radar position, and whose bottom is quadrilateral, they can be detected. Through servos, the target is maintained at center position of the cruciform area. That is to track the target.

When the radar beam aims at a target, the reflecting signals are received by the antenna of the guidance station, and then sent to the receiver. The signals are amplified, and mixed to be converted into intermediate frequency signals. The signals are filtered, and detected to form the video signals. The video signals are amplified and supplied to displays and the coordinate tracking system. The coordinate tracking system yields the signals, which represent coordinates of the target, and send them to the command generator and the computer of impact distance. According to the target coordinates and motion parameters, the impact distance from the guidance station is estimated for the commander to choose launching opportunity.

When the missile enters the desired volume of space, where the command antenna beam illuminates, the command transmitter transmits interrogating pulses. The pulses trigger the missile transponder. The transponder transmits responding signals to the guidance station immediately. The pitch antenna and azimuth antenna can receive the responding signals and reflecting signals of the target. By using frequency selection method, the responding signals and target signals are separated. The missile signals are fed to the missile receiver. Similarly to the target, the missile receiver also provides the video signals about slant range, pitch angle, and azimuth angle of the missile to the displays and the coordinate tracking system. The coordinate tracking system yields the signals, which represent coordinates of the missile, and sends them to the command generator.

If the guidance station guides three missiles, then the command generator can produce flight control commands k_1 and k_2, and one time command k_3 freeing from the third safety of the fuse respectively for three missiles. These command voltages are transformed into command pulses, encoded to be transmitted to missiles through command antenna along with interrogating pulses.

The commands k_1 and k_2 received by the command receiver on board are properly processed to be fed to autopilot, which controls flight trajectory. When the missile-target distance reach a certain value, the one time command k_3, which is received by the on-board receiver, opens the third safety of the fuse in order to make the fuse in a state of readiness before detonation. When the distance between target and missile reaches effective range of the missile warhead, the fuse detonates the warhead to damage the target.

9.1.2 Actual Flight Phases

In general, for a command guidance missile, its actual trajectory is not completely coincident with its kinematical trajectory, but undulates weakly around its kinematical trajectory, finally approaches to be in line. See Figure 9.1 – 3.

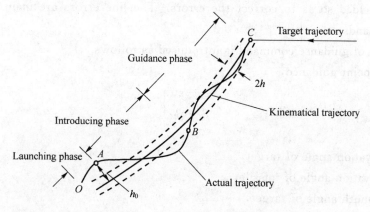

Figure 9.1 – 3 Trajectory of a command guidance missile

For a command guidance missile, as to properties of the actual trajectory, an intercepting course is approximately divided into three phases:

① Launching phase. When the missile leaves from the launching shelf, as the initial speed of the missile is small, and the operating efficiency of the aerodynamic fins is low, the missile will not be controled. The time of the uncontrolled flight is very short. After the transitorily uncontrolled flight phase, the autopilot begins to stabilize or control the attitude of the missile to ensure the missile attitude or position to reach requirement. For the missile with the booster, when the booster is working, in general, only the roll channel is controlled. After the booster is cast, the pitch control channel and yaw control channel are switched on. The location of the missile at the launching phase end distributes randomly. Referring to Figure 9.1 – 3, the distance h_0 from A point to the kinematical trajectory is initial error. The initial error sometimes reaches several hundreds of meters.

② Introducing phase. From A point, the missile begins to receive the control commands, the error gradually decreases. The undulant motion is evident. When the missile enters into the desired area, which is determined by two dashed curves at h distance from the kinematical trajectory, the phase ends. The phase corresponds to the AB segment.

③ Guidance phase. The guidance commands guide the missile to fly, and to intercept the target. The trajectory is relatively smooth. The phase corresponds to the BC segment.

9.1.3 Command Generation

The guidance course means that based on line errors, the guidance commands are continuously yielded so as to correct the errors. The line errors are main components of guidance commands.

Basic term of guidance command is introduced as follows.

For three-point guidance,

$$\left. \begin{array}{l} \varepsilon_T = \varepsilon_M \\ \beta_T = \beta_M \end{array} \right\}$$

where

ε_T = elevation angle of target,

ε_M = elevation angle of missile,

β_T = azimuth angle of target,

β_M = azimuth angle of missile.

In actual flight, there are errors, namely

$$\left. \begin{array}{l} \Delta\varepsilon = \varepsilon_T - \varepsilon_M \\ \Delta\beta = \beta_T - \beta_M \end{array} \right\} \qquad (9.1-1)$$

Usually use the form of line error to denote guidance commands. After transforming the angle errors into the line errors,

$$\left. \begin{array}{l} h_{\Delta\varepsilon} = \Delta\varepsilon R_M \\ h_{\Delta\beta} = \Delta\beta R_M \end{array} \right\} \qquad (9.1-2)$$

where R_M is the slant range of the missile.

In Reference [2], the line errors are

$$\left. \begin{array}{l} h_{\Delta\varepsilon} = \Delta\varepsilon R_M \\ h_{\Delta\beta} = \Delta\beta R_M \cos\varepsilon_M \end{array} \right\} \qquad (9.1-3)$$

For half-lead angle guidance, the guidance equations (7.2-25) and (7.2-26) correspond to line errors as follows:

$$\left. \begin{array}{l} h_\varepsilon = h_{\Delta\varepsilon} - \dfrac{1}{2} \dfrac{\dot{\varepsilon}_T}{\dot{\Delta r}} \Delta r R_M \\ \\ h_\beta = h_{\Delta\beta} - \dfrac{1}{2} \dfrac{\dot{\beta}_T \cos\varepsilon_T}{\dot{\Delta r}} \Delta r R_M \end{array} \right\} \qquad (9.1-4)$$

9.1 Principle of Command Guidance

As the width of the wave beam is generally narrow, for instance 10°, if the lead quantity is not restricted, the missile will be able to move out of the wave beam. If required lead angle is less than 5°, the maximum of $|\Delta r|$ and the minimum of $|\Delta \dot{r}|$ have to be limited. In order to confine maximum of $|\Delta r|$, the following limiting law should be used:

$$\Delta r_L = \Delta r \left(1 - \frac{\Delta r}{R_0}\right) \quad (9.1-5)$$

where R_0 is a maximum. Figure 9.1-4 shows Δr_L versus Δr curve. In order to confine minimum of $|\Delta \dot{r}|$, the following limiting law should be used.

$$\Delta \dot{r}_L = \begin{cases} -a, & \text{if } |\Delta \dot{r}| \leqslant a \\ \Delta \dot{r}, & \text{if } |\Delta \dot{r}| > a \end{cases} \quad (9.1-6)$$

where a is a constant, such as 700 m/s. Figure 9.1-5 indicates $\Delta \dot{r}_L$ versus $\Delta \dot{r}$ curve.

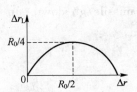

Figure 9.1-4 Δr_L versus Δr curve **Figure 9.1-5** $\Delta \dot{r}_L$ versus $\Delta \dot{r}$ curve

Thus,

$$\left. \begin{aligned} h_\varepsilon &= h_{\Delta\varepsilon} - \frac{1}{2} \frac{\dot{\varepsilon}_T}{\Delta \dot{r}_L} \Delta r_L R_M \\ h_\beta &= h_{\Delta\beta} - \frac{1}{2} \frac{\dot{\beta}_T \cos \varepsilon_T}{\Delta \dot{r}_L} \Delta r_L R_M \end{aligned} \right\} \quad (9.1-7)$$

In a practical system, in order to improve dynamical performance, it is necessary to change the basic term, such as introducing lead compensation, or PD (proportional plus derivative) compensation, or PID compensation, as well as, limiting magnitude. For example, for SAM—2 missile, the lead compensator is applied as follows:

$$\overline{h}_\varepsilon + \frac{T_1 s}{1 + T_2 s} h_\varepsilon, \quad \overline{h}_\beta + \frac{T_1 s}{1 + T_2 s} h_\beta \quad (9.1-8)$$

where $T_1 \gg T_2$,

$$\overline{h}_\varepsilon = \begin{cases} h_\varepsilon, & \text{if } |h_\varepsilon| \leqslant 175 \text{ m} \\ 175 + \frac{1}{6}(h_\varepsilon - 175), & \text{if } h_\varepsilon > 175 \text{ m} \\ -175 + \frac{1}{6}(h_\varepsilon + 175), & \text{if } h_\varepsilon < -175 \text{ m} \end{cases}$$

For another example, Sidewinder-type missile uses PD compensation.

In order to reduce dynamical error, the following dynamical compensation should be added.

$$\left.\begin{aligned} h_{D\varepsilon} &= X(t)\dot{\varepsilon}_T \\ h_{D\beta} &= X(t)\dot{\beta}_T \cos\varepsilon_T \end{aligned}\right\} \qquad (9.1-9)$$

For half-lead angle method, since the curvature of the kinematical trajectory is small, $X(t)$ is taken constant. For three-point method, the curvature of the kinematical trajectory is greater than that of half-lead angle method, and furthermore, the nearer the missile approaches the target, the greater the kinematical trajectory curvature will be, so $X(t)$ is taken a function proportional to time, i.e.,

$$X(t) = X_0 + Kt \qquad (9.1-10)$$

The gravity always holds downward to make the missile go down. In view of the disadvantage, the following gravity compensation is added,

$$h_{g\varepsilon} = \frac{g\cos\gamma}{K_0} \qquad (9.1-11)$$

where

γ = flight path angle,

K_0 = open loop gain of guidance loop.

The gravity has effect only in vertical plane. So the gravity compensation is added only to the pitch channel.

For SAM—2 missile, the elevation and azimuth commands are

$$\begin{bmatrix} \lambda_\varepsilon \\ \lambda_\beta \end{bmatrix} = \begin{bmatrix} \left(\bar{h}_\varepsilon + \dfrac{T_1 s}{1+T_2 s} h_\varepsilon + h_{D\varepsilon} + h_{g\varepsilon}\right)\dfrac{1+T_3 s}{1+T_4 s} \\ \left(\bar{h}_\beta + \dfrac{T_1 s}{1+T_2 s} h_\beta + h_{D\beta}\right)\dfrac{1+T_3 s}{1+T_4 s} \end{bmatrix} \qquad (9.1-12)$$

where $T_4 > T_3$, $\dfrac{1+T_3 s}{1+T_4 s}$ is lag compensator. Since the lead compensator is introduced, the system frequency band becomes wide. As a result, this can increases the undulate disturbance. In order to increase open loop gain, and limit the width of the frequency band, the lag compensator is introduced.

λ_ε and λ_β are created in the observing coordinate system, but the missile implements control in missile coordinate system. If the missile employs the configuration of "+" fins and stabilizes roll-attitude, the observing coordinate system is approximately consistent with the performing coordinate system of the missile. If the missile employs the configuration of "X" fins, the difference between the two coordinate systems is 45° as shown in Figure 9.1-6.

Here, the guidance intelligence λ_ϵ, λ_β are projected onto the missile coordinate system as shown in Figure 9.1-7.

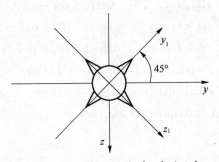 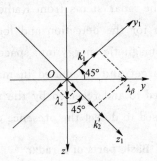

Figure 9.1-6 Geometrical relation between both coordinate systems

Figure 9.1-7 The guidance intelligence projection in the performing system

k_1' is an algebraic sum of the projections of λ_ϵ and λ_β along y_1 axis. k_2' is an algebraic sum of the projections of λ_ϵ and λ_β along z_1 axis. From Figure 9.1-7,

$$\left. \begin{aligned} k_1' &= -\lambda_\epsilon \sin 45° + \lambda_\beta \cos 45° \\ k_2' &= \lambda_\epsilon \cos 45° + \lambda_\beta \sin 45° \end{aligned} \right\} \tag{9.1-13}$$

The missile hardly rotates about the longitudinal axis because of the roll autopilot on board, while the tracking radar rotates all the time for tracking the target. From Figure 7.2-2, it can be seen that the projection $\dot{\beta}_T \sin \epsilon_T$ causes the observing axis x of Figure 7.2-2 to spin, as a result, not to retain the 45° difference between them, but to have a torsional angle φ. φ is

$$\varphi = \int_{t_0}^{t} \dot{\beta}_T \sin \epsilon_T \, dt \tag{9.1-14}$$

So the control commands are improved below:

$$\left. \begin{aligned} k_1 &= -\lambda_\epsilon \sin(45° - \varphi) + \lambda_\beta \cos(45° - \varphi) \\ k_2 &= \lambda_\epsilon \cos(45° - \varphi) + \lambda_\beta \sin(45° - \varphi) \end{aligned} \right\} \tag{9.1-15}$$

9.2 Guidance Stations

Radar is main equipment of the guidance station. Before discussing the guidance station, basic conceptions of radar are first introduced.

9.2.1 Basic Concepts of Radars

The radar stems from Radio Detection and Ranging. The radar is an electromagnetic system for the detection and location of reflecting objects or targets. The radar radiates electromagnetic energy into space by an antenna. Some of the radiated energy is intercepted by a target and reradiated in many directions. A portion of reradiated (echo) energy is returned to and received by the radar antenna, which delivers it to a receiver. There it is processed to detect the presence of the target and determine its location.

1. Basic parts of a radar

Figure 9.2 – 1 shows a block diagram of a conventional pulse radar.

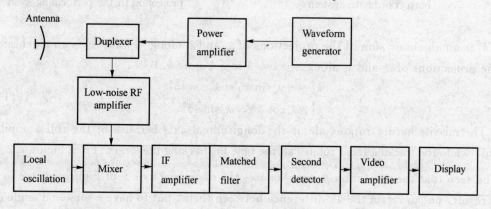

Figure 9.2 – 1 Block diagram of a conventional pulse radar

The radar antenna is what connects with the radar to the outside world. It serves the following functions: ① concentrates the radiated energy on transmitting; ② collects the echo energy from the target on receiving; ③ measures the angle of arrival of the received echo signal so as to provide the location of a target in azimuth, elevation, or both; ④ acts as a spatial filter to resolve (separate) targets in angle (spatial) domain; and ⑤ allows the desired volume of space to be observed.

The duplexer allows a single antenna to be used on a time-shared basis for both transmitting and receiving. It acts as a rapid switch to protect the receiver from burning out on transmitting; it directs the echo signal to the receiver rather than to the transmitter on receiving.

The transmitter may be a power amplifier, such as the klystron, traveling wave tube, or transistor amplifier. A modulator turns the transmitter on and off in synchronism with the input pulses. The transmitter may also be a power oscillator, such as the magnetron. The magnetron oscillation has been widely used for pulse radars of modest capability. Since the magnetron provides relatively low average power (one or two kilowatts) and poor stability, the amplifier is preferred for applications requiring long-range detection of small moving targets in the presence of large clutter echoes. When a power oscillation is employed, it is also turned on and off by a pulse modulator to generate a pulse waveform. Because the radar signal is produced at low power by a waveform generator before being delivered to the power amplifier, it is far easier to achieve the special waveforms needed for pulse compression and for coherent systems such as moving-target indication (MTI) radar and pulse doppler radar.

The receiver is almost always a superheterodyne. The mixer of the superheterodyne receiver and local oscillator translate the RF signal to an intermediate frequency. It is amplified by the IF amplifier. The IF amplifier also includes the function of the matched filter; one which maximizes the output signal-to-noise ratio. Maximizing the signal-to-noise ratio at the output of the IF maximizes the detectability of the signal. The second detector is an envelope detector which eliminates the IF carrier and passes the modulation envelope. When doppler processing is employed, as it is in CW (continuous-wave), and pulse doppler radars, the envelope detector is replaced by a phase detector which extracts the doppler frequency by comparison with a reference signal at the transmitted frequency. There must also be included filters for rejecting the stationary clutter and passing the doppler-frequency-shifted signals from moving targets. The video amplifier raises the signal power to a level where it is convenient to display the information it contains. As long as the video bandwidth is not less than half of the IF bandwidth, there is no adverse effect on signal detectability.

A low-noise receiver front end (the first stage) is desirable for many civil applications, but in military radars the lowest noise figure attainable might not always be appropriate. In a high-noise environment of unintentional interference or hostile jamming, the radar with a low-noise receiver is more susceptible than one with higher noise figure. In addition, a low-noise amplifier as the front end generally will result in the receiver having less dynamic range. In order to avoid the disadvantages of a low-noise-figure receiver, the low-noise RF amplifier stage is sometimes omitted and the mixer is used as the receiver front end. The higher noise figure of the mixer can then be compensated by an equivalent increase in the transmitter power.

CHAPTER 9   Command Guidance Systems

A threshold is established at the output of the video amplifier to allow the detection decision to be made. The decision is based on the magnitude of the receiver output. If the receiver output crosses the threshold, a target is said to be present. If it does not cross the threshold, only noise is assumed to be present. The decision may be made by an operator, or it might be done with an automatic detector without operator intervention.

The purpose of signal processor is to reject undesired signals (such as clutter, noise) and pass desired echo signals. It is performed prior to the threshold detector where the detection decision is made. Signal processing includes the matched filter and the doppler filters.

Some radars process the detected target signal further, in the data processor, before displaying the information to an operator. An example is an automatic tracker.

The signal processor and data processor are usually implemented with digital technology.

2. Radar equation

The power density at range R from a radar that radiates a power P_t from an antenna of gain G_t is

$$\frac{P_t G_t}{4\pi R^2} \tag{9.2-1}$$

The reradiated power density back at the radar is

$$\frac{P_t G_t}{4\pi R^2} \frac{\sigma}{4\pi R^2} \tag{9.2-2}$$

where $\sigma=$ radar cross section of a target.

The received signal power P_r of the radar is

$$P_r = \frac{P_t G_t}{4\pi R^2} \frac{\sigma}{4\pi R^2} A_e \tag{9.2-3}$$

where $A_e=$ effective area of receiving antenna.

When the received signal power P_r just equals the minimum detectable signal S_{min}, the maximum range of a radar R_{max} occurs. The radar equation may be written as

$$R_{max} = \left[\frac{P_t G_t A_e \sigma}{(4\pi)^2 S_{min}}\right]^{1/4} \tag{9.2-4}$$

In view of $G_t = \dfrac{4\pi A_e}{\lambda^2}$, where $\lambda=$ wavelength of radar signal,

$$R_{\max}=\left(\frac{P_t A_e^2 \sigma}{4\pi\lambda^2 S_{\min}}\right)^{1/4} \qquad (9.2-5)$$

or

$$R_{\max}=\left[\frac{P_t G_t^2 \lambda^2 \sigma}{(4\pi)^3 S_{\min}}\right]^{1/4} \qquad (9.2-6)$$

The simple form of the radar equation is instructive, but not very useful because it leaves out many things.

The minimum detectable signal is

$$S_{\min}=kT_0 BF_n\left(\frac{S}{N}\right)_{\min} \qquad (9.2-7)$$

where

$k=$ Boltzman's constant $=1.38\times 10^{-23}$ J/K,

$T_0=290$ K,

$B=$ receiver bandwidth,

$F_n=$ noise figure,

$\left(\dfrac{S}{N}\right)_{\min}=$ minimum detectable signal-to-noise ratio.

The product kT_0 is equal to 4×10^{-21} J.

Since the radar waveform is a repetitive series rectangle-like pulse, the numerator of the radar equation is multiplied by a factor $nE_i(n)$, where $E_i(n)$ is the efficiency in adding together n pulses. The power P_r is the peak power of a radar pulse. The average power P_{av} is $P_{av}=P_t\tau f_p$, where $f_p=$ pulse repetition frequency, and $\tau=$ pulse duration. Considering the effect of the environment on the propagation of radar waves, the radar equation should be multiplied by the propagation factor F^4. In addition, considering the system losses (which is defined as the system loss factor L_s), the radar equation (9.2-4) becomes

$$R_{\max}=\left[\frac{P_{av}G_t A_e \sigma n E_i(n) F^4}{(4\pi)^2 kT_0 F_n (B\tau) f_p (S/N)_{\min} L_s}\right]^{1/4} \qquad (9.2-8)$$

In most radar designs the product $B\tau\approx 1$, the equation (9.2-8) becomes

$$R_{\max}=\left[\frac{P_{av}G_t A_e \sigma n E_i(n) F^4}{(4\pi)^2 kT_0 F_n f_p (S/N)_{\min} L_s}\right]^{1/4} \qquad (9.2-9)$$

In general, for each particular application, conventional radar equation has to be tailor to that specific application.

3. Radar frequencies

Before the powerful electromagnetic wave signal produced by the radar transmitter is modulated, the signal frequency is called radar frequency. Radars can operate at frequencies as low as 2 MHz (just above AM the broadcast band) and as high as several hundred GHz (millimeter wave region). More usually, radar frequencies may be from about 5 MHz to over 95 GHz. For military secrecy, during World War II, letter-bands (such as S, X and L) were used to designate the frequency bands; but the letter designations were continued to be used after the war. They have been officially accepted as an IEEE Standard. Table 9.2-1 lists the letter-band designations.

Table 9.2-1　IEEE Standard letter designation for radar-frequency bands

Band designation	Nominal frequency range	Band designation	Nominal frequency range
HF	3~30 MHz	K_u	12~18 GHz
VHF	30~300 MHz	K	18~27 GHz
UHF	300~1 000 MHz	K_a	27~40 GHz
L	1~2 GHz	V	40~75 GHz
S	2~4 GHz	W	75~110 GHz
C	4~8 GHz	mm	110~300 GHz
X	8~12 GHz		

Radar characteristics vary with frequency band. In general, at the lower frequencies it is easier to obtain high-power transmitter and physically large antenna. Thus, it is easier to achieve long range. On the other hand, at the higher frequency, it is easier to achieve accurate measurements of range and location because the higher frequencies provide wider bandwidth (which determines range accuracy and range resolution) as well as narrower beam antennas for a given physical size antenna (which determines angle accuracy and angle resolution).

4. Antenna radiation pattern

The distribution of electromagnetic energy in three-dimensional angular space, when plotted on a relative (normalized) basis, is called the antenna radiation pattern. This distribution can be plotted in various ways, e.g., polar or rectangular, voltage intensity or power density, or power per unit solid angle (radiation intensity). The radiation pattern varies with the range, space angle (elevation angle, azimuth angel). Figure 9.2-2 is a typical radiation pattern. The maximum lobe of the pattern is the main lobe (main beam). The main lobe is surrounded by minor lobes (also sidelobes). The minor lobe approximately

180° from the main lobe is called the back lobe.

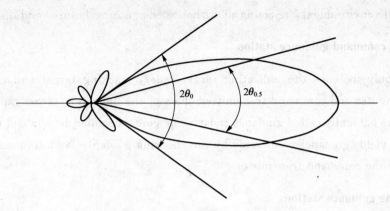

Figure 9.2-2 Antenna radiation pattern

The null beamwidth is an angle between both zero power radiation directions of the main lobe in a pitch plane or azimuth plane. The $2\theta_0$ in Figure 9.2-2 is the null beamwidth. The most frequently expressed width is the half-power beamwidth (HPBW), which occurs at the 0.707-relative-voltage-level, or 0.5-relative-power-level, or the 3 dB level. The $2\theta_{0.5}$ in Figure 9.2-2 is the half-power beamwidth. The half-power beamwidth is given by

$$\text{HPBW} = KD/\lambda \tag{9.2-10}$$

where

  D is the aperture dimension,
  λ is the wavelength,
  K is a proportionality constant.

9.2.2 Types of Guidance Stations

A guidance station is referred to as an off-board guidance system as well. It acquires the intelligence about the missile and target, guides the missile to fly at the target. In the light of the site of the guidance stations, the guidance stations are classified into truck-carrier stations, ship-carrier stations, and airplane-carrier stations. According to work modes, the guidance stations are divided into remote control (command, beam-riding) stations, homing guidance stations, and combined guidance stations. The guidance stations have functions of searching target, identifying target, evaluating menace, pre-electing target, sorting order, acquiring and tracking missile, yielding and transmitting guidance commands, transmitting

CHAPTER 9 Command Guidance Systems

target and missile information to commanding center, detecting and analyzing electromagnetic environment, rejecting disturbance, diagnosing failures and displaying.

1. Radio command guidance station

It commonly includes the indicating target radar, tracking target and missile radar, command computer and command transmitter. Except the indicating target radar, the other equipments are generally called guidance radar. The guidance radar detects and tracks targets and missiles, yields guidance commands by the choosing guidance law, transmits commands to missiles by the command transmitter.

2. Homing guidance stations

In the light of guidance modes, homing guidance stations are classified into the passive homing guidance station, active homing guidance station, and semi-active homing guidance station. The passive homing guidance stations are simpler. One passive homing guidance station may be an optical sighting device. By eyeballing, one can determine the target position. If a target enters killing volume of space, then a shooter launches a missile. The active homing guidance station consists of a searching and tracking radar, and computers. It searches and tracks targets, and provides data for choosing launch opportunity. After being launched, the missile is guided by an active homing seeker instead of the commands from the guidance station. The semi-active guidance station comprises a searching radar, a tracking radar, an illuminating radar, and computers. The illuminating radar provides energy of illuminating target. The semi-active radar seeker on board receives reflecting signal from the target, and yields guidance commands to guide flight of the missile.

American Patriot missile and Russian C—300 missile use command and TVM (Track via Missile) combined guidance. The Patriot missile is equipped with a C-band phased-array radar, AN/MPQ53. The radar is responsible for searching, identifying, illuminating, tracking, and guiding. Its reaching coverage ranges from: 0° to 90° in elevation direction, and from $-60°$ to $+60°$ in azimuth direction. Its detecting range is 160 kilometers. It can track 8 missiles, and guide 8 missiles simultaneously.

9.2.3　Radars of Guidance Stations

1. Search radar

The functions of the search radar are to detect targets in defense space, find targets as fast as possible, continuously measure targets data, identify targets, assign and indicate targets, and do surveillance for combat. For attacking the targets, it assigns the target data to the tracking radar. In this case, the tracking radar rapidly captures and tracks the targets to shorten reactive time of combat.

The search radar may be an independent radar, or a sort of work mode of multi-function phased-array radar. The tasks of searching and indicating targets are accomplished by a single radar, a few radars, or early warning net of aerial defense system. For example, American improved Hawk missile is equipped with two radars. One of them is a pulse indication radar, AN/MPQ50, which detects high altitude targets. The other is a continuous wave indication radar, AN/MPQ48, which detects low altitude targets.

2. Illuminating radar

The illuminating radar offers the illuminating energy to target and forward-wave signal to the missile for semi-active homing missile such that the seeker gets the angular velocity of the target relative to the missile. Sometimes, through forward-wave signal, transmits demanded control commands, such as, changing proportional navigation constant, opening fuse, or delaying of fuse. In order to detect the Doppler frequency shift, which is proportional to the velocity of the target relative to the missile, the seeker needs the coherent reference signal of the illuminating radar, so except the front antenna in the seeker, the rear antenna is used to receive forward-wave signal. The illuminating radar may be a continuous wave, or pulse wave system. For example, American improved Hawk missile is equipped with a continuous wave illuminating radar, AN/MPQ−39.

3. Tracking and guiding radar

The tasks of such a radar is to track the missile and target, accurately measure their positions, and guide the missile flight at the target. For example, the guidance radar of the Soviet C−25 ground to air missile employs mechanical-electronic scan system. In $10°\times10°$ space it may track the target and missile, and measure their coordinates, whose accuracy reaches $1' \sim 2'$. The Sidewinder-type missile applies monopulse tracking radar. In addition, in initial phase, the infrared tracker guides the missile to enter the radar beam, and then shifts to radar tracker. The radical type of Rapier missile uses the optical sighting device, while the improved type adds monopulse tracking radar. American Patriot missile and Russian C—300 missile adopt multi-function phrase-array radar for tracking.

9.2.4 Guidance Radar Systems

According to the technique characteristics of the tracking radar, the guidance radar systems include scan (conical scan, linear scan) systems, monopulse systems, and phased array systems.

In conical scan, let the axis of the antenna beam offset a squint angle from the axis of rotation, and rotate at definite angular rate (see Figure 9.2-3). If the target locates in the reference axis, the echo signal amplitude does not vary with the beam rotation, and retains a constant. If the target deviates from the reference axis, the echo signal amplitude varies periodically with the beam rotation as shown in Figure 9.2-4. The amplitude of the modulation is proportional to the angular distance between the target direction and the rotation axis. The phase of the modulation determines the target direction.

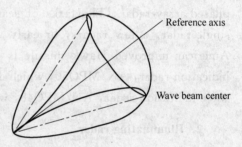

Figure 9.2-3 Conical scan

The received pulse train is held at an average amplitude U_0 by the AGC loop, operating with a large time constant. The instantaneous pulse amplitude, on a steady target near the axis, is

$$U(t) = U_0[1 + k_s \varepsilon \cos(\omega_s t - \phi)] \qquad (9.2-11)$$

where

$\varepsilon =$ the magnitude of the error angle,

$\phi =$ the direction of the error with respect to the horizontal,

$\omega_s =$ the radian scan rate ($\omega_s = 2\pi f_s$),

$k_s =$ the error-slop factor of the antenna.

In normal operation, the error will not exceed a small fraction of the beamwidth, and the error slop will be constant which can be evaluated on the tracking axis. If the error is resolved into two separate components, the equation (9.2-11) may be written as

$$U(t) = U_0[1 + k_s\varepsilon_t\cos\omega_s t + k_s\varepsilon_e\sin\omega_s] \qquad (9.2-12)$$

where ε_t and ε_e are the traverse and elevation errors, respectively.

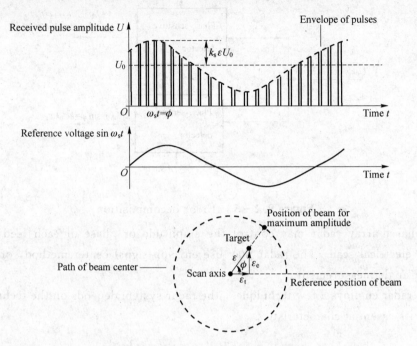

Figure 9.2-4 Error detection in conical scan radar

An envelop detector can sense the scan-rate components of signal modulation and reject the DC level and components at the repetition rate and above. As a result, the error signal $U_0 k_s \varepsilon \cos(\omega_s t - \phi)$ is obtained. The error signal is compared with two orthogonal reference signals in phase-sensitive detectors to generate $kU_0 k_s \varepsilon_t$ and $kU_0 k_s \varepsilon_e$ as shown in Figure 9.2-5.

A secant correction is necessary in any conventional elevation-over-azimuth tracking radar where the elevation drive system rotates when the antenna changes azimuth. That is, the azimuth control signal is $kU_0 k_s \varepsilon_t \sec\varepsilon$, where ε is the antenna axis elevation angle. Two error signals are amplified to drive the antenna servo motors of elevation and azimuth.

The linear scan radar is also called fan-shaped scan radar. The scanning angle from initial position is proportional to the scanning time. It makes use of envelop signal center method for angle measurement.

The monopulse radar gets two dimension angle coordinates in the light of one echo pulse.

CHAPTER 9 Command Guidance Systems

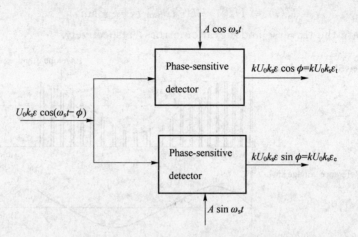

Figure 9.2-5 Error decomposition

The phased-array radar may control the amplitude or phase of each feed element to implement electrical scan. The radar may use envelop signal center method, or monopulse method for angle measurement.

When radar employs a few techniques, the radar system depends on the technique which represents its essential characteristics.

9.3 Linear Scan Radar

The linear scan radar linearly scans along narrow wave beam direction at fan-shaped tabular wave beam. The echo from a target is modulated by the antenna radiation pattern to form a bell shape envelope. The time from initial scan point to the envelope center corresponds to the target angle. The radar uses the track-while-scan mode.

9.3.1 Angle Measurement

The linear scan radar can track target and missile, and measure their angles. In order to measure the elevation angle and azimuth angle, two orthogonal antennas are used to scan. This scan can form "+" space area (see Figure 9.1-2). The radar can measure absolute angles in ground coordinate system, and relative angles relative to the initial scan position.

1. Tracking target in angle

The linear scan radar moves two gates such that the two gates are centered on the envelope for angle measurement. The scan angle from initial position is proportional to the scan time. The scan time interval T_t from initial position to the envelope center represents scanning angle. The scanning angle is

$$\theta_t = \frac{\theta}{T - T_r} T_t = \omega T_t \qquad (9.3-1)$$

where

$\theta=$ scanning range,
$T=$ wave beam scan period,
$T_r=$ wave beam return time,
$\omega=$ wave beam scan angular velocity.

Target tracking in angle may be done manually by an operator who manually captures the target and turns the antenna to maintain a marker on the display over the target echo pip. Manual track is an auxiliary track mode. It may be used to create the condition for automatic track.

An angle tracker can track a target in angle, and generate the angle pulse, which represents the elevation angle coordinate ε_T, or azimuth angle coordinate β_T.

Figure 9.3-1 indicates a block diagram of an automatic angle tracker. It includes a time discriminator, an error amplifier, an integrator, a comparer, and a gate generator.

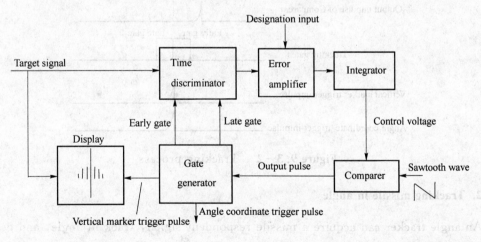

Figure 9.3-1 Block diagram of automatic track

The control voltage, which is dependent on tracking error, and the sawtooth wave voltage are compared in the comparer to generate the output pulse as shown in Figure 9.3-2. That is, if the two voltages are equal, the comparer produces the output pulse. It can make tracking gates (i.e. early and late gates) to move. The gate generator gives two gates: early and late gates. In addition, it generates vertical marker trigger pulse for display and angle coordinate trigger pulse for angle measurement. There are fixed time delays between these pulses and the output pulse of the comparer. In automatic track case, according to the error, which the center line of tracking gates deviate from the center line of the target echo envelope, the time discriminator yields the error signal. The error signal is amplified, and integrated to provide the control voltage. It can make tracking gates be centered on the target echo envelope.

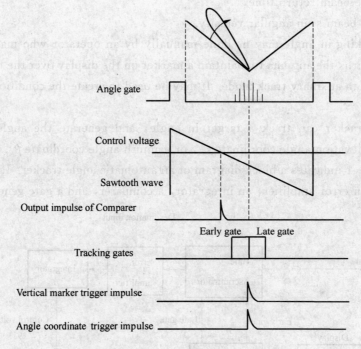

Figure 9.3 – 2 Tracking process

2. Tracking missile in angle

An angle tracker can acquire a missile respondent signal, track in angle, and generate the angle pulse, which represents the elevation angle coordinate ε_M, or azimuth angle coordinate β_M.

In principle, the angle measurement for the missile is similar to the angle measurement for the target. In linear scan system, in order to obtain a target echo signal, the antenna needs two times modulations: on transmission and reception, while for the missile respondent signal the antenna needs only one time modulation on reception. This produces the error of measuring angle. Furthermore, the greater the range, the greater the error will be. It needs to be properly compensated. The angle track method for the missile is also the gate track method. The process of the angle measurement for the missile is different from that of angle measurement for the target. For the missile, the angle tracker has waiting, wide gate tracking, and narrow gate tracking states.

The waiting state is set for acquiring the missile. In waiting state, wider tracking gates are employed, and the center of tracking gates is located at the center of beam scan time. Once the missile respondent signal enters tracking radar beam, the wider tracking gates will reliably capture the respondent signal. Before the missile respondent signal is acquired, the missile is at the initial phase, and not controlled by the guidance radar. During this phase, the guidance radar continuously tracks the target. It is possible for the missile not to drop into the beam center because there is initial dispersion. It is difficult to capture the missile respondent signal via using narrower track gates (corresponding to 3° space angle), but it is easy to capture the missile respondent signal via using wider tracking gates (corresponding to 20° space angle). When the wider tracking gates are tracking steadily in angle, the wider tracking gates shift to the narrower tracking gates for tracking in angle.

9.3.2 Range-tracking Systems

A range-tracking system is used to measure the range to a target. Range measurement boils down to measuring time delay between the transmission of an RF pulse and the echo signal from the target. The slant range R from the guidance station to the target is

$$R = \frac{1}{2} ct \tag{9.3-2}$$

where

$t =$ time delay,

$c =$ velocity of propagation of electromagnetic wave, 3×10^8 m/s.

1. Tracking target in range

In the early days of radar, target tracking in range was usually done manually by an operator who watched a display and positioned handwheel or handcrank to maintain a marker on the display over the target echo pip. The time delay between both the range pulse, which coincides with the marker pulse, and reference pulse represents the slant range. For manual tracking, tracking accuracy depends on the operator's skill. Because an operator continuously operates for tracking a target, the work inflicts physical tiredness and psychological burden on the operator.

A range tracker can track a target in range, and generate the range pulse, which represents the range R_T. Figure 9.3-3 indicates a block diagram of an automatic range tracker. It includes a time discriminator, an error amplifier, an integrator, a variable delay circuit, and a pulse generator.

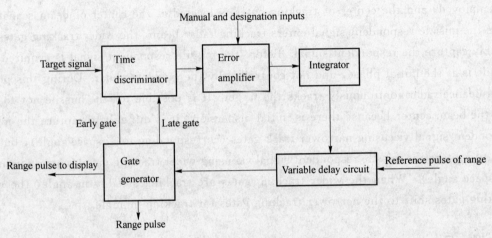

Figure 9.3-3 Block diagram of automatic range tracker

The technique for automatic tracking in range is based on the split gate. Referring to Figure 9.3-4, two range gates are generated. One is the early gate and the other is the late gate. The error of the range gates relative to the echo pulse center of the target is sensed, and the circuitry is provided to respond to the error voltage by moving the gates in a direction to re-center on the echo pulse center. In Figure 9.3-4, the portion of the signal in the early gate is less than that of the late gate. The signals in the two gates are integrated and subtracted to produce the difference error signal. The sign of the difference determines the direction of the movement of the pair of range gates in order to have the pair straddle the echo pulse. The amplitude of the difference determines how far the pair of gates should be moved to the center of the pulse. When the error signal is zero, the range gates are centered on the pulse and the position of the two gates gives the target's range. Deviation of the pair of gates from the center of the echo pulse increases the signal energy in one of the gates and decreases it in the other. The error signal causes the two pulses to be moved so as to reestablish the equilibrium.

9.3　Linear Scan Radar

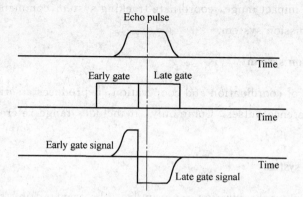

Figure 9.3-4　Split-gate range tracking

If the track gates deviate from the target echo envelope center, the time discriminator produces the error signal. The error is amplified to be fed to the integrator. The integrator yields control voltage to change the width of the output of the variable delay circuit. This makes the track gates move correspondingly to be centered on the echo signal. The range pulse represents the range of the target.

2. Tracking missile in range

A range tracker can acquire a missile respondent signal, track in range, and generate the range pulse representing the range R_M.

A missile range tracking system has waiting state and automatic track state. Before the missile is launched, for waiting the tracking gates of the missile range are set a time delay, which corresponds to 2.1 kilometers. Once the missile travel reaches the range, the tracking gates capture the missile respondent signal rapidly, and then the tracking system shifts to automatic track state.

9.3.3　Components of Linear Scan Guidance Radar

The linear scan guidance radar, which acts as the guidance equipment outside the missile for the command guidance system, in the light of functions, generally includes three component parts: tracking device, command generator, and command transmission device. But, an actual guidance radar is very complex. For example, a type of linear scan guidance radar includes ten systems: synchronization system, transmitting system, antenna-feed system, receiving system, displaying system, antenna and launcher control system, computing system of impact range, coordinate tracking system, command generation system, and command transmission system.

1. Synchronization system

For the purpose of coordination and cooperation, it produces a series of synchronization pulses, or called reference pulses. Commonly, it includes range reference pulses and angle reference pulses.

2. Transmitting system

Triggered by the synchronization range pulse, the transmitting system generates and transmits the powerful radio frequency signal (high frequency detecting pulses). The radio frequency energy generated is radiated to space by way of the antenna-feed system.

3. Antenna-feed system

It is to transmit and radiate the powerful radio frequency signal (high frequency detecting pulses), high frequency interrogating pulses, and commands produced by the command transmitter. In addition, it receives the echo of the target and the responding signal of the missile.

4. Receiving system

It has two receiving channels, which respectively receive the target echo and the respondent signal of the missile. According to frequency difference between them, they are separated by the high frequency pre-selector, and amplified, then transformed to become video signals. The video signals are supplied to the coordinate tracking system and the display system.

5. Displaying system

In linear scan guidance radar, there are several displays. They are used to display the space positions of the target and missile. For example, an intensity-modulated rectangular display indicates elevation (or azimuth) angle in horizontal direction, and rang in vertical direction. By watching displaying information, operators choose launching opportunity, manually track target, or evaluates attacking results.

9.3 Linear Scan Radar

6. Antenna and launcher control system

The function of an antenna control system is to control the antenna rotations in both horizontal and elevation directions to implement search, and track for the target. The launcher isochronously rotates with the antenna to ensure the launcher to aim at the target, and the missile to fall into the antenna beam area in favor of acquiring the missile rapidly.

The antennas have their respective servos to perform closed-loop control.

7. Computing system of impact range

Missile-target impact range from the guidance station is estimated so that the commander can choose launching opportunity.

There is a simple approximate formula as follows:

$$R_E = \frac{R_T V_M}{\dot{R}_T + K V_M} \quad (9.3-3)$$

where

R_E = range of the impact point from the guidance station,
R_T = radial range of target,
V_M = average velocity of missile,
K = constant, $K < 1$.

8. Coordinate tracking system

It receives the video signals of the target and missile from the receiving system to generate their range and angle pulses to be fed to the command generation system.

9. Command generation system

It produces the pitch control command k_1 and yaw control command k_2 of missile flight, the one time command k_3 opening the third safety of the radio fuse. For example, in the case of no-jam, if $\Delta R = R_T - R_M = 435$ m, where R_T and R_M represent the ranges of the target and the missile respectively, the one time command k_3 is sent out. In the case of jam and no measuring range, when $t = 21$ s, the one time command k_3 is sent out.

Generating three commands needs input signals as shown in Figure 9.3-5.

In Figure 9.3-5,

- ε_0 = reference pulse of elevation angle,
- β_0 = reference pulse of azimuth angle,

CHAPTER 9 Command Guidance Systems

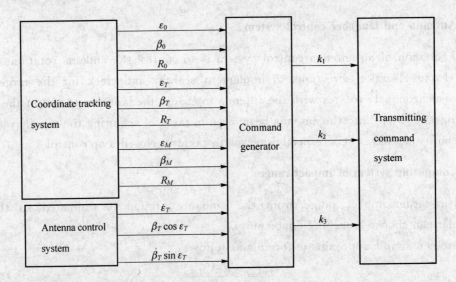

Figure 9.3-5 Input-output relation of command generation system

- $R_0 =$ synchronization pulse of range,
- $\varepsilon_T =$ elevation angle pulse of target,
- $\beta_T =$ azimuth angle pulse of target,
- $R_T =$ range coordinate pulse of target,
- $\varepsilon_M =$ elevation angle pulse of missile,
- $\beta_M =$ azimuth angle pulse of missile,
- $R_M =$ range coordinate pulse of missile.

The time interval between the angle coordinate pulse and the reference pulse represents the angle. The time relation between the range coordinate pulse and the synchronization pulse of range represents the range.

$\dot{\varepsilon}_T$, $\dot{\beta}_T \cos \varepsilon_T$ and $\dot{\beta}_T \sin \varepsilon_T$ are provided by the antenna control system.

k_1 and k_2 are voltage signals changing slowly. k_3 is a constant voltage signal.

10. Command transmission system

The control command voltages, k_1 and k_2, a constant voltage signal k_3 are transformed to pulses. By way of encoding, through the command antenna, in the light of definite time order, they and interrogating pulse are radiated together to a missile.

9.4 Commands Transmission

9.4.1 Transmission Channel

A principle block diagram of command transmission is indicated in Figure 9.4-1.

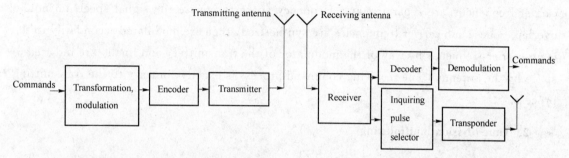

Figure 9.4-1 Principle block diagram of command transmission

The command transmission channel includes two parts. One part consists of a modulator, an encoder, a command transmitter, and a transmitting antenna in ground guidance station. The other part consists of a receiving antenna, a decoder, and an onboard demodulator. In addition, when the guidance station transmits interrogating signal, the transponder of missile transmits respondent signal to the guidance station for tracking and measuring missile.

9.4.2 Command Types

In command guidance system, the guidance station transmits guidance commands to each missile, such as the control command of yaw channel, the control command of pitch channel, the one time command, and the interrogating signal. Furthermore a guidance station may generally control multi-missiles. For example, a typical guidance station needs to guide three missiles, and transmit 11 signals:

① Control commands of three missiles, $k_{1-1}, k_{2-1}, k_{1-2}, k_{2-2}, k_{1-3}, k_{2-3}$,
② One time command of three missiles, $k_{3-1}, k_{3-2}, k_{3-3}$,
③ Interrogating pulses, R_i,
④ Reference pulses of dividing extent, T.

Transmitting 11 commands relates to the multiplex problem.

CHAPTER 9 Command Guidance Systems

9.4.3 Multiplex Manners of Commands

The following multiplex manners are generally used.

1. Frequency-division multiplexing

The commands are modulated to respective sub-carrier frequencies. The respective sub-carrier frequencies are separated at a frequency space to ensure the signal spectrum not to overlap. These sub-carrier frequencies are synthesized, then are modulated secondarily to the main carrier frequency by way of the modulator of the transmitter, and finally are radiated to space by the antenna. The receiving command process is exactly contrary to the transmitting process.

2. Time-division multiplexing

The control commands are assigned at respective time intervals for transmitting. For example, a command transmission period is divided into some equal time intervals, which are called dividing extent, frame, and row respectively. Respective commands are arranged at the chosen dividing extent, frame, and row. On schedule, they are transmitted. In order to ensure the synchronization relation of the transmitting signal and receiving signal, the reference pulse of dividing extent is needed to be transmitted for synchronization of the receiving system. They will be discussed in details later.

9.4.4 Modulation of Commands

The control commands k_1, k_2, which are produced by the command generator, are slow changing continuous signals. They can not be directly transmitted to space. Therefore it is necessary to convert them into pulses. A conversion method is the phase time modulation, which expresses the magnitude and polarity of the command voltage with the relative location of a pulse.

The command voltage and sawtooth wave are sent to the time modulation (voltage comparer). The period of the sawtooth wave is equal to that of the dividing extent (reference pulse), If both of them are equal, the voltage comparer produces the command pulse. The command position relative to the reference pulse represents the magnitude and polarity of the command voltage as shown in Figure 9.4-2.

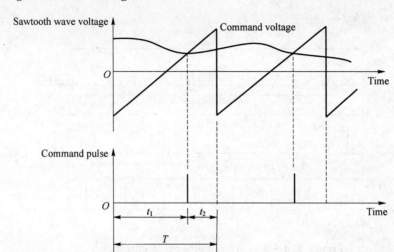

Figure 9.4-2 Pulse time modulation

Relative command value is

$$k = \frac{u_k}{u_{k,\max}} = \frac{t_1 - t_2}{T} \tag{9.4-1}$$

where

$u_k =$ instantaneous value of command voltage,

$u_{k,\max} =$ maximum of command voltage,

$T =$ Period of dividing extent.

If the command voltage is zero, the corresponding command pulse is located at the center. If the command voltage is greater than zero, the corresponding command pulse is located at the right half side. Contrarily, it is located at the left half side.

9.4.5 Transmission Time Arrangement of Command Pulses

The time arrangement of command pulses is needed to ensure commands not to overlap, and moreover to reflect the magnitude change of the commands. The time division is used for the arrangement. There are a lot of interspaces between the reference pulses, so the commands are inserted.

For example, 1 second is divided into 44 equal parts, one of which is called a dividing extent, i.e., reference time interval T. Every dividing extent is divided into 64 equal parts, one of which is a frame. Every frame is divided into 8 equal parts, one of which is called a row. See Figure 9.4-3.

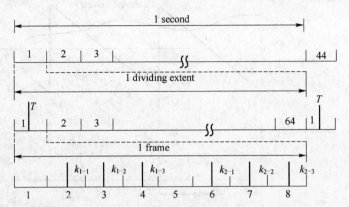

Figure 9.4-3 Arrangement of command pulse

The reference pulse of the dividing extent is located at the 5th row of the 1st frame. The 33rd frame is a middle frame. The zero voltage is located at the 5th row of the 33rd frame.

k_{1-1}, k_{1-2}, k_{1-3}, k_{2-1}, k_{2-2}, k_{2-3} (i.e. the 3rd missile k_2) are located respectively at the 2nd row, 3rd row, 4th row, 6th row, 7th row, and 8th row. If the command voltage changes, the position of the command will jump from the row of original frame to the corresponding row of another frame. i.e., the frame number changes, while the row number is unchanged. For example, the command pulse jumps a frame when the command voltage changes 2 V. If the command voltage of k_{1-1} is -20 V, 0 V, 20 V respectively, the position of the command pulse is located at the 2nd row of the 23rd frame, the 2nd row of the 33rd frame, the 2nd row of the 43rd frame respectively. It may be found that the frame number, which the command pulse locates at, represents the magnitude and polarity of the command voltage, and the row number represents the command attribute. Every command has a fixed row number, but the command frame number changes with the magnitude and polarity.

One time commands have on-off states, and have no magnitude and polarity. The 2nd, 3rd, 4th, 6th, 7th, and 8th rows are occupied. The 5th row may be occupied by the dividing extent pulse. So the one time commands occupy the leftover 1st row. However one time commands of different missiles need to be differentiated, so they are distributed in different frames. For instance, k_{3-1} is located at the 1st row of 5th, 21st, 37th, or 53rd frame, k_{3-2} is located at the 1st row of 9th, 25th, 41st, or 57th frame, and k_{3-3} is located at the 1st row of 13th, 29th, 45th, or 61st frame.

The frequency of interrogating pulse is one ninth of the row repetition frequency. The interrogating pulse appears one time every 9 rows, and at any row of any frame. In order to avoid confusion, the interrogating pulses are arranged at early time of every row, while the command pulses are arranged at late time of every row as shown in Figure 9.4 – 4.

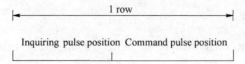

Figure 9.4 – 4 Relation between the interrogating pulse and command pulse

9.4.6 Command Pulse Encoding

In order to guarantee every missile to receive the control command which belongs to itself, the pulse encoding is used. For example, a triple encoding group can differentiate different command pulses through different characteristic parameters, τ_1 and τ_2. Figure 9.4 – 5 illustrates triple encoding groups of k_1 commands.

Time delays τ_1 and τ_2 of every missile command pulse and diving extent are different. According to the differences, receivers of missiles differentiate respective command pulses and dividing pulses.

When the interrogating pulse is not encoded, the difference between the interrogating pulse and the command pulse depends on pulse width. The width of the interrogating pulse is one half of the width of the command pulse. The width discriminator on board may differentiate them.

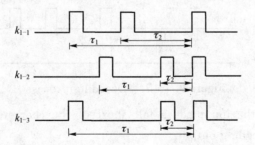

Figure 9.4 – 5 Triple encoding groups of k_1

These triple encoding groups and interrogating pulses are modulated to high frequency pulses for transmission.

9.4.7 Command Decoding and Demodulation

1. Command decoding

Decoding purpose is to differentiate encoding groups. The principle of decoder is shown in Figure 9.4 – 6.

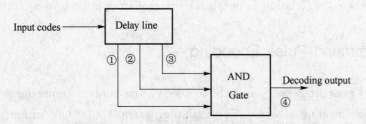

Figure 9.4 – 6 Principle of decoder

After the pulses of input codes are delayed by delay line, the delay line produces outputs ①, ② and ③. Signal ① is original signal, not being delayed; signal ② is delayed for τ_2; signal ③ is delayed for τ_1. The AND gate transforms the three signals into decoding pulse ④ as shown in Figure 9.4 – 7.

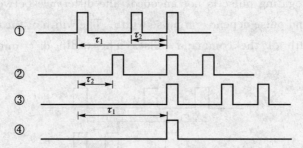

Figure 9.4 – 7 Decoding process

Apparently, when the input codes fall short of the characteristic parameters of the decoder, there is not decoding output.

2. Command demodulation

The task of command demodulation is to restore the command pulses to the command voltages. The command pulses of pulse time modulation are demodulated to the pulse duration modulation. By way of filter, obtain the command voltages. Demodulation principle of command pulses is shown in Figure 9.4 - 8. The signal of command voltage is sent to autopilot.

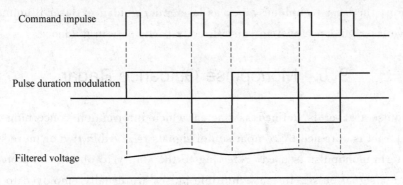

Figure 9.4 - 8 Demodulation principle of command pulse

9.5 Brief Introduction to Optical-electronic Technique

The electronic-jam, radar stealthy technique etc. put forward challenges for radar guidance. Because the optical systems have some advantages, they are applied to guidance weapon systems. A lot of guidance stations are equipped with visible light device, infrared device, or laser device. They act as aided guidance means, or even participate in entire guidance process. For instance, the Sidewinder-type missile system uses the infrared system to lead the missile into radar track beam in initial phase.

Compared with the radar (radio frequency) systems, the optical systems possess advantages as follows:

① It may directly obtain visible images, or infrared images, so it is easy to identify target,

② There is no effect of electronic-jam,

③ Visible light wave length and infrared wave length are shorter, so their resolutions are higher,

④ Low altitude performance is good.

Certainly, the optical systems have the following deficiencies:

① Operating range is shorter (nearer),

② It is susceptible to weather condition.

The optical systems, which are commonly applied, are

① visible light image track system,

② infrared image track system,
③ laser ranging system.

In addition, the laser techniques, such as laser-rider guidance, laser illuminating source for semi-active laser guidance system, also directly participate in guidance.

9.6 Monopulse Guidance Radar

A monopulse tracker is defined as one in which information concerning the angular location of a target is obtained by comparison of signals received in two or more simultaneous beams. The term monopulse is used, referring to the ability to obtain complete angle error information on a signal pulse. In fact, multiple pulses are usually employed to increase the probability of detection, and improve the accuracy of the angle measurement. Due to utilizing the signals which appear simultaneously in more than one beam for an angle measurement, in the accuracy of the measurement, the monopulse tracker is superior to time-shared single-beam tracking systems (such as conical scan) which suffer degradation when the echo signal amplitude changes with time. Hence, the monopulse systems eliminate the effects of echo amplitude fluctuations of the target. The monopulse technique has an inherent capability for high precision angle measurement because its feed structure is compact with short signal paths and rigidly mounted with no moving parts.

Amplitude-comparison monopulse and phase-comparison monopulse are used for a monopulse angle measurement. The most widely used monopulse is amplitude-comparison monopulse which compares the amplitudes of the signals simultaneously received in multiple squinted beams to determine the angle.

9.6.1 Amplitude-comparison Monopulse

Two overlapping antenna patterns with their main beams pointed in slightly different directions are employed as shown in Figure 9.6-1. They may be generated by using two feeds slightly displaced in opposite directions from the focus of a parabolic reflector. If a target locates in OL direction, both signals received in two beams are equal in amplitudes. The OL line is called equisignal line. Any deviation of the target off the OL line can result in the two signals unequal. In order to perform amplitude-comparison technique, hybrid junctions are used. The rat-race, or hybrid-ring junction is a type of hybrid junction. The hybrid-ring junction provides the sum and difference signals of the two input signals. The difference signal determines the magnitude of the angle error, while the phase difference between the difference signal and the sum signal provides the target direction relative to the equisignal line. The magnitude of the difference signal is proportional to the deviation angle from the equisignal axis. The sum signal also provides range measurement, as well as acts as a reference for determining the sign of the angle measurement.

Figure 9.6-2 shows a hybrid-ring junction. There are four arms around the half circle. The path between each two adjacent arms is odd number times of $\lambda/4$.

If the signals are inputted at arms PA and PB, the signal at arm PΣ is sum signal because the path from arm PA, or arm PB to arm PΣ is $\lambda/4$ (i.e. equal path), where λ is the wavelength, that is, at arm PΣ, the two signals are in phase, and the signal at arm PΔ is the difference signal because the difference of two paths is $3\lambda/4-\lambda/4=\lambda/2$, that is, at arm P$\Delta$, two signals are out of phase 180°.

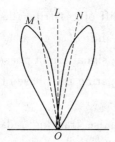

Figure 9.6-1 Monopulse antenna patterns

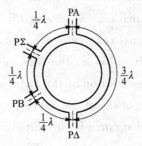

Figure 9.6-2 Rat-race or hybrid-ring junction

In the same manner, if the arm PΣ acts as the input, it is divided equally in power and appears with the same phase at both arm PA and arm PB. Nothing will appear at arm PΔ. The two signals are fed to two feed-horns. By way of the parabolic reflector, they are radiated to space.

If a target is centered, the two horns receive equal energy signals, which are sent to the arm PA and arm PB of the hybrid-ring junction. At the arm PΣ we obtain the sum signal

$$u_\Sigma = u_A + u_B \qquad (9.6-1)$$

And at the arm PΔ we obtain the difference singal

$$u_\Delta = u_A - u_B = 0 \qquad (9.6-2)$$

If the target moves off the equisignal axis, there is an unbalance of energy in the horns, that is, the signal at the arm PA is different from the signal at arm PB. For example, the signal at arm PA is greater than the signal at arm PB. The arm PΣ provides the sum signal with the phase shift of $\lambda/4$ trip relative to the arm PA or arm PB, and the arm PΔ provides the difference signal. The amplitude of the difference signal determines the offset angle of the target. The phase of the difference signal depends on the arm PA signal. Because the phase shift of the arm PΔ signal relative to arm PA signal is the phase shift of $\frac{3}{4}\lambda$ trip, the phase shift of the arm PΔ signal relative to the arm PΣ signal is a phase shift of $\frac{3}{4}\lambda - \frac{1}{4}\lambda = \frac{1}{2}\lambda$ trip, i. e., π phase shift. That is, the sum signal and the difference signal are out of phase.

If the target deviate in the other direction, the signal at arm PB is greater than the signal at arm PA. In the same manner, the arm PΣ provides the sum signal with the phase shift of $\lambda/4$ trip relative to the arm PA or arm PB, and the arm PΔ provides the difference signal. The phase of the difference signal depends on the arm PB signal. Because the phase shift of the arm PΔ signal relative to arm PB signal is the phase shift of $\lambda/4$ trip, the phase shift of the arm PΔ signal relative to the arm PΣ signal is a phase shift of $\lambda/4 - \lambda/4 = 0$ trip, i. e., null phase shift. That is, the sum signal and the difference signal are in phase.

Thus, the direction of the angle error is found by comparing the phase of the difference signal with the phase of the sum signal.

A block diagram of the amplitude-comparison monopulse tracking radar for a single angular coordinate is shown in Figure 9.6 - 3. The two adjacent antenna feeds (i. e. feedhorns) are connected to the two input arms of a hybrid junction. When two signals (such as the signals from the squinted beams) are inserted at the two input ports, the sum and difference of the two are obtained at the output arms. On reception, the output of the sum and difference ports are each heterodyned to an intermediate frequency and amplified in the superheterodyne receiver. In amplitude-comparison monopulse, it is necessary that the sum and difference channels have the same phase and amplitude characteristics. For this reason, a single local oscillator is shared by the two channels. Through a duplexer (TR), the transmitter is connected to the sum port of the hybrid junction. The TR is one that switches off the receiver (if the receiver is not switched off, the receiver is damaged because the transmitter is powerful.) when transmitting, and switches on the receiver after transmitting. The automatic agin control (AGC) is used for maintaining the received signal amplitude stable. The outputs of the sum and difference channels are delivered to a phase-sensitive detector, which is a nonlinear device that compares two signals of the same frequency. The sum signal at the IF output provides a reference signal to the phase-sensitive detector, which derives angle-tracking-error

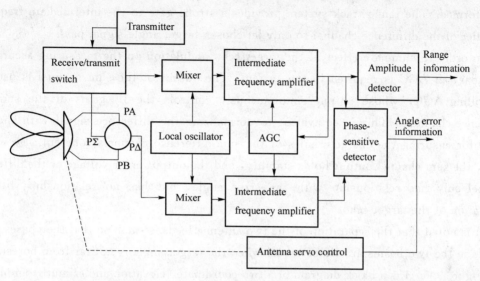

Figure 9.6-3 Block diagram of the amplitude-comparison monopulse tracking radar of a single angular coordinate

voltage from the difference signal. The phase-sensitive detector is essentially a dot-product device producing the output voltage

$$u_e = \frac{|\Sigma|}{|\Sigma|} \frac{|\Delta|}{|\Sigma|} \cos\phi \quad \text{or} \quad u_e = \frac{|\Delta|}{|\Sigma|} \cos\phi \qquad (9.6-3)$$

where

$|\Sigma|$ = magnitude of sum signal,

$|\Delta|$ = magnitude of difference signal,

ϕ = phase difference between difference and sum signals, as mentioned early, $\phi=0$, or $\phi=\pi$.

Thus,

$$u_e = \begin{cases} \dfrac{|\Delta|}{|\Sigma|}, & \text{if } \phi=0 \\ -\dfrac{|\Delta|}{|\Sigma|}, & \text{if } \phi=\pi \end{cases} \qquad (9.6-4)$$

The magnitude of the output u_e of the phase-sensitive detector is proportional to $|\theta_T - \theta_0|$, where θ_T = target angle and θ_0 = boresight angle relative to a reference direction. The sign of the output u_e of the phase-sensitive detector indicates the direction of the angle error relative to the boresight. The error voltage passes through a moderate low-pass filter to the servo amplifier so as to correct the antenna direction. Before the angle tracking, the range tracking

CHAPTER 9 Command Guidance Systems

is performed. The range track system provides a strobe gate to the intermediate frequency amplifier of the difference channel to only let chosen target angle signal pass.

In order to eliminate effect of the target echo undulation on the measuring accuracy of the radar, an AGC is required in the system. The output of the sum channel is used for controlling AGC, whose output simultaneously controls the IF gains of the sum and difference channels. This is equivalent to normalizing the difference signal with the sum signal for ensuring the characteristics of the sum-difference channels coherent. By way of AGC, the sum channel output holds stability, and the output error voltage of the difference channel only has relationship with the error angle, but has no relationship with the undulation of the target echo.

It is noted that the separation of the two antenna feeds is small so that the phases of the signals in the two beams are almost equal when the target angle is not far from boresight.

Figure 9.6-4 is a block diagram of a two-coordinate (elevation and azimuth) amplitude-comparison monopulse tracking radar. The cluster of four feed horns generates four partially overlapping beams. The four feeds may be used to illuminate a parabolic reflector, such as Cassegrain reflector. The arrangement of the four feeds is shown in the mid left-hand portion of the figure. The four feeds are connected to the four hybrid junctions.

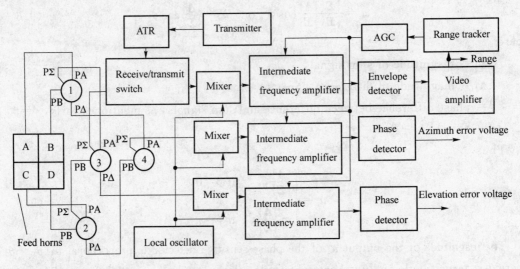

ATR—Antitransmit-receive

Figure 9.6-4 Block diagram of two-coordinate (elevation and azimuth) amplitude-comparison monopulse tracking radar

On transmission, the detecting signals are fed to the arm PΣ of the hybrid-ring junction 3. The symmetric arms PA, PB of the hybrid-ring junction 3 send the same phase and same amplitude signals to the arms PΣ of the hybrid-ring junctions 1 and 2. By way of the symmetric arms PA and PB of the hybrid-ring junctions 1 and 2, the detecting signals are fed to the feed horns A, B, C, and D, and then by way of parabolic antenna, are radiated to space. Here, the hybrid-ring junctions 1, 2, and 3 act as the power dividers.

On reception, all four feeds are used to generate the sum pattern. The difference pattern in one plane is formed by taking the sum of two adjacent feeds and subtracting them from the sum of the other two adjacent feeds. The difference pattern in the orthogonal plane is obtained similarly. For example, based on the arrangement of the feeds shown in Figure 9.6-4, the sum pattern is found from $S_A + S_B + S_C + S_D$; the azimuth difference pattern is obtained from $(S_A + S_C) - (S_B + S_D)$; the elevation difference pattern is found from $(S_A + S_B) - (S_C + S_D)$. Note that the upper feeds form the lower beams when radiated by a reflector antenna. Referring to Figure 9.6-4, the sum signal $S_A + S_B + S_C + S_D$ is obtained at the arm PΣ of the hybrid-ring junction 3; the difference signal $(S_A + S_B) - (S_C + S_D)$ (i.e. the elevation difference signal) is obtained at the arm PΔ of the hybrid-ring junction 3; and the difference signal $(S_A + S_C) - (S_B + S_D) = (S_A - S_B) + (S_C - S_D)$ (i.e. the azimuth difference signal) is obtained at the arm PΣ of the hybrid-ring junction 4. The three mixers for the sum, elevation difference, and azimuth difference channels share a common local oscillation to better maintain the phase relationships among the three channels. Two phase-sensitive detectors extract the angle-error information; one for azimuth and the other for elevation. Range information is extracted from the output of the sum channel after envelope detection.

9.6.2 Phase-comparison Monopulse

Figure 9.6-5 shows a phase-comparison monopulse. Although two antenna beams are also used to obtain angle measurement in the phase-comparison monopulse, the two beams look in the same direction and cover the same region of space rather than be squinted to look in two slightly different directions. For this reason, two antennas have to be employed in the phase-comparison monopulse rather than using two feeds at the focus of a single antenna as is the case for an amplitude-comparison monopulse. The amplitudes of the signals are the same, but their phases are different. This is just the opposite of

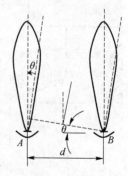

Figure 9.6-5 Phase-comparison monopulse in one angle coordinate

CHAPTER 9   Command Guidance Systems

the amplitude-comparison.

Consider two antennas spaced a distance d apart, as in Figure 9.6-5. If the signal arrives from a direction θ with respect to the normal to the baseline, the phase difference of the signals received in the two antennas is

$$\varphi = \varphi_A - \varphi_B = \frac{2\pi}{\lambda} d \sin \theta \tag{9.6-5}$$

where λ = wavelength. The phase difference arises from the wave-way difference. A measurement of the phase difference of the signals received in the two antennas can provide the angle θ to the target. The phase-comparison monopulse is sometimes called an interferometer radar.

Figure 9.6-6 shows a block diagram of the phase-comparison monopulse in one angle. Angle information can also be extracted in a phase-comparison monopulse by using sum and difference patterns and processing the signals similar to that described for the amplitude-comparison method. Because there is a phase shift of 90° between the sum and difference signals as shown in Figure 9.6-7, a 90° phase shift has to be introduced in the difference signal so that the output of the phase-sensitive detector is an error signal whose amplitude is a function of the sine of the angle of arrival from the target measured with respect to the perpendicular to the two antennas.

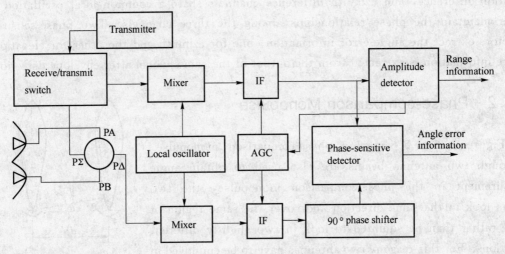

Figure 9.6-6   Block diagram of the phase-comparison monopulse in one angle

One of the limitations of phase-comparison monopulse is the effect of grating lobes due to the separation d of the two antennas each of dimension d. If the spacing d between the

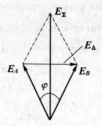

Figure 9.6 – 7 Vector diagram of sum and difference signals of phase-comparison monopulse

phase centers of the antenna is greater than that of the antenna diameter, high sidelobes are generated in the sum pattern and ambiguities can occur in the angle measurement. Even when the spacing is the same as the antenna diameter, a poor antenna pattern can also appear. In practice, the separation between the two antennas should be less than the antenna diameter d if good radiation patterns are to be obtained on transmission and angle ambiguities are to be avoided on reception. Relative to the more popular amplitude-comparison method, the phase-comparison monopulse method is seldom used.

9.7 Phased-array Radar

The phased-array radar is one in which phased array antenna is used to perform the beam scan in space. Its antenna aperture retains stationary, i.e. without mechanical motion, but it relies on electronic means to perform changes of the antenna orientation. The antenna comprises a lot of radiating elements, which are arranged to array in light of chosen law. By the beam control computer, one may change the phase relation and amplitude relation between the antenna radiation elements, quickly change the antenna orientation and the beam shape, and obtain the antenna-aperture illumination function corresponding to needed antenna pattern. In general, there are three scanning manners: phase scanning, time-delay scanning, and frequency scanning. The phase scanning is widely used, so the scan manner will be discussed as follows.

9.7.1 Principle of Phase Scanning

The beam of an antenna points to a direction that is normal to phase front. In phased arrays, this phase front is adjusted to steer the beam by controlling the excitation phase of

CHAPTER 9 Command Guidance Systems

each radiating element. In practice, the phase scanning is to control the phase shift quantity of the phase shifter (each antenna element has a phase shifter) to perform beam scanning.

A line array consists of N antenna elements equally spaced a distance d apart as shown in Figure 9.7 – 1. The elements are assumed to be isotropic radiators. The phase shift quantities of each phase shifter are $0, \alpha, 2\alpha, \cdots, (N-1)\alpha$ respectively.

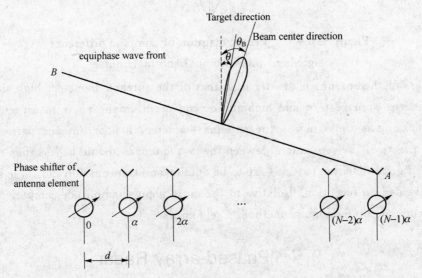

Figure 9.7 – 1 One dimensional line array antenna

A point field intensity of deviation θ with respect to the normal line in far space is a vector sum of each antenna element radiation field, i.e.

$$\boldsymbol{E}(\theta) = \boldsymbol{E}_0 + \boldsymbol{E}_1 + \cdots + \boldsymbol{E}_{N-1} = \sum_{k=0}^{N-1} \boldsymbol{E}_k \qquad (9.7-1)$$

Because the range from the antenna to the target is long, the effect which the micro-difference in range has on the amplitude will be neglected. In view of the same amplitude feed, the amplitudes of producing radiation field of each element are considered to be equal, denoted by E. If the element 0 is taken as the reference with zero phase, then

$$\boldsymbol{E}(\theta) = |\boldsymbol{E}| \sum_{k=0}^{N-1} e^{jk\left(\frac{2\pi}{\lambda}d\sin\theta - \alpha\right)} \qquad (9.7-2)$$

where $\alpha = \frac{2\pi}{\lambda} d \sin \theta_B$, $\theta_B =$ offset angle of beam center.

Let $\varphi = \frac{2\pi}{\lambda} d \sin \theta$, i.e. the wave-way difference gives rise to the phase difference between adjacent elements. $k\alpha$ is a phase shift of phase shifter k relative to phase shifter 0.

Let $X = \frac{2\pi}{\lambda} d \sin\theta - \alpha$, and apply geometrical series formula to obtain

$$E(\theta) = |E| \frac{1 - e^{jNX}}{1 - e^{jX}} \qquad (9.7-3)$$

Use Euler's formula to obtain

$$E(\theta) = |E| \left(\frac{\sin \frac{N}{2} X}{\sin \frac{1}{2} X} e^{j\frac{N-1}{2}X} \right) \qquad (9.7-4)$$

If $X = \frac{2\pi}{\lambda} d \sin\theta - \alpha = 0$, each component is in phase, and the field intensity amplitude is maximum, i.e.

$$|E(\theta)|_{max} = NE$$

Thus, normalized field-intensity pattern function is

$$F(\theta) = \frac{|E(\theta)|}{|E(\theta)|_{max}} = \left| \frac{1}{N} \frac{\sin \frac{N}{2} X}{\sin \frac{1}{2} X} \right| \qquad (9.7-5)$$

When $X = 0$, $F(\theta)$ is maximum 1, and the beam is centered on the target (i.e. $\theta = \theta_B$). Thus,

$$\frac{2\pi}{\lambda} d \sin\theta - \alpha = 0, \quad \text{i.e.} \quad \frac{2\pi}{\lambda} d \sin\theta_B - \alpha = 0$$

$$\sin\theta_B = \frac{\lambda}{2\pi d} \alpha \qquad (9.7-6)$$

$$\theta_B = \arcsin\left(\frac{\lambda}{2\pi d} \alpha\right) \qquad (9.7-7)$$

Equation (9.7-6) shows changing adjacent elements' phase difference α, which the phase shifter supplies, may change the antenna beam center direction. As a result, this forms beam scanning.

In Figure 9.7-1, the electromagnetic wave phase of each point in AB line is equal, and AB line is known as equiphase front. The direction of the pattern maximum is perpendicular to the equiphase front.

Equation (9.7-5) is rewritten as follows

$$F(\theta) = \left| \frac{1}{N} \frac{\sin \frac{N}{2} X}{\sin \frac{1}{2} X} \right| = \left| \frac{1}{N} \frac{\sin\left[\frac{N\pi}{\lambda} d (\sin\theta - \sin\theta_B)\right]}{\sin\left[\frac{\pi}{\lambda} d (\sin\theta - \sin\theta_B)\right]} \right| \qquad (9.7-8)$$

If $\frac{\pi}{\lambda} d (\sin\theta - \sin\theta_B) = 0, \pm\pi, \pm 2\pi, \cdots, n\pi$ (n is integer), the maximums of $F(\theta)$ occur.

$\frac{\pi}{\lambda}d(\sin\theta-\sin\theta_B)=0$ corresponds to the main lobe. The other maximums correspond to grating lobes (See Figure 9.7-2). The grating lobe appearance leads to multi-value of angle measurement. In order to avoid the appearance of grating lobes, ensure

$$\left|\frac{\pi}{\lambda}d(\sin\theta-\sin\theta_B)\right|<\pi$$

or

$$\frac{d}{\lambda}<\frac{1}{|\sin\theta-\sin\theta_B|}$$

Because of $|\sin\theta-\sin\theta_B|\leqslant 1+|\sin\theta_B|$, the condition without grating lobes is

$$\frac{d}{\lambda}<\frac{1}{1+|\sin\theta_B|} \qquad (9.7-9)$$

Because the beam scan coverage is unfavorably too great, people generally take $|\theta_B|\leqslant 60°$, or $|\theta_B|\leqslant 45°$, corresponding to $\frac{d}{\lambda}<0.53$ or $\frac{d}{\lambda}<0.59$.

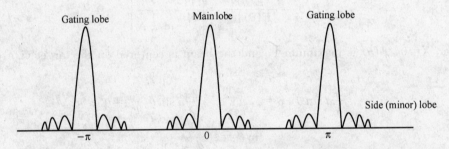

Figure 9.7-2 Main lobe and gating lobes arise in pattern

In general, element number N is large, but X is small, the equation (9.7-8) becomes

$$F(\theta)=\left|\frac{\sin\frac{N\pi}{\lambda}d(\sin\theta-\sin\theta_B)}{\frac{N\pi}{\lambda}d(\sin\theta-\sin\theta_B)}\right| \qquad (9.7-10)$$

For $\frac{\sin x}{x}$, if $x=1.39$, $\frac{\sin x}{x}=\frac{1}{\sqrt{2}}$. Therefore, obtain half-power beamwidth at pattern. Thus,

$$\frac{N\pi}{\lambda}d(\sin\theta-\sin\theta_B)=1.39 \qquad (9.7-11)$$

Let $\theta=\theta_B+\frac{1}{2}\Delta\theta_{1/2}$, where $\Delta\theta_{1/2}$ is half-power beamwidth. In view of $\sin\theta=\sin\left(\theta_B+\frac{1}{2}\Delta\theta_{1/2}\right)\approx\sin\theta_B+\cos\theta_B\frac{1}{2}\Delta\theta_{1/2}$, substituting this equation into equation (9.7-11) produces

$$\Delta\theta_{1/2} \approx \frac{1}{\cos\theta_B} \times \frac{0.88\lambda}{Nd} \quad (\text{rad}) \qquad (9.7-12)$$

i. e.

$$\Delta\theta_{1/2} \approx \frac{1}{\cos\theta_B} \times \frac{51\lambda}{Nd} \quad (°) \qquad (9.7-13)$$

If $d=\lambda/2$, then

$$\Delta\theta_{1/2} \approx \frac{1}{\cos\theta_B} \times \frac{102}{N} \qquad (9.7-14)$$

The equation (9.7 - 14) shows that the half-power beamwidth is inversely proportional to the cosine of θ_B, i. e. the greater the scan angle θ_B, the wider the half-power beamwidth will be.

In the θ_B direction, the gain of the antenna is

$$G(\theta_B) = \frac{4\pi A_e}{\lambda^2} = \frac{4\pi N_0 d^2}{\lambda^2}\cos\theta_B \qquad (9.7-15)$$

where, $A_e =$ effective aperture area, i. e. projection of the antenna area $A=N_0 d^2$ ($N_0 =$ total number of array elements) onto the equiphase plane. The increase in the scan angle θ_B leads to the decrease in the antenna gain. Thus, with the increase in θ_B, the beam widens, but the antenna gain decreases. Therefore, the beam scan coverage is confined in $\pm 60°$ or $\pm 45°$.

Except one-dimensional line arrays, there are two dimensional planar arrays, multi-surface arrays, and dome arrays etc. The planar arrays are widely used. The planar array has two configurations: rectangle configuration, triangle configuration. Similar to scan principle of the line array, the planar array can scan in two-dimensional space. Please refer to relevant literature about detail discussion.

9.7.2 Space Feed

From the transmitter to each antenna element, there is a feed network (beamformer) for power division; from each antenna element to the receiver, there is a feed network for power addition. The track radars in Russian C—300 missile weapon system and American Patriot missile weapon system use space feed. The space feed is also known as optical feed. There are two types of the space feeds: lens array as shown in Figure 9.7 - 3 and reflectarray as shown in Figure 9.7 - 4.

The transmitting process for lens array is that the radio frequency energy, which is radiated by the feed, is collected by the array elements facing to the feeds, passes through the phase shifter for shifting phase, and then is sent to the opposite array elements. Finally they radiate beam into space. The receiving process is contrary to the transmitting process. According to proper law changing the phase of each phase shifter makes the antenna beam perform scanning in light of requirement. If necessary, the transmitting and receiving tasks may be separately implemented by two feeds. There is an angle α separation between both feeds as shown in Figure 9.7 - 3. In this case, the phase compensation is needed for the wave-way difference between the receiving and transmitting.

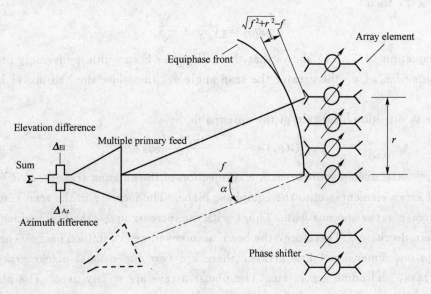

Figure 9.7 - 3 Lens array

For the spherical phase front, a phase correction is necessary, and given by

$$\frac{2\pi}{\lambda}(\sqrt{f^2+r^2}-f) \qquad (9.7-16)$$

The transmitting process for reflectarray is that the radio frequency energy from the offset feed, enters the antenna elements, passes through the phase shifts, is reflected, and passes back through the phase shifters to be radiated as a plane wave in the desired direction. Because the energy passes through each phase shifter twice, a phase shifter needs only be capable of half the phase shift needed for a lens array or a conventional array. The phase shifter must be reciprocal so that there is a net controllable phase shift after passing through the device in both directions. This limitation rules out frequently used ferrite or garnet phase shifters. To avoid aperture blocking, the primary feed may be offset as shown in Figure 9.7 - 4. As before, transmitting and receiving feeds may be separated.

9.7 Phased-array Radar

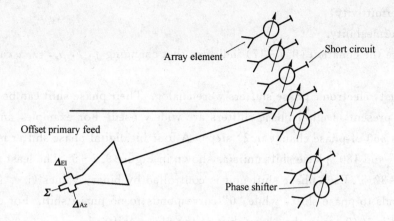

Figure 9.7-4 Reflectarray

9.7.3 Phase Shifters

Phase shifters are also known as phasors. There are a lot of phase shifters in phased array radar. The phase shifters are key devices for scan. So phase shifters should possess following properties in various degrees: ability to change phase rapidly (a few microsecond), capability of handling high peak and high average power, little drive power, low loss, insensitivity to changes in temperature, small size, low weight, and low cost. Phase shifters are classified into two categories: reciprocal and non reciprocal. For the reciprocal phase shifters, the phase shift in one direction (i.e. transmit) is the same as the phase shift in the opposite direction (i.e. receive). All diode phase shifters are reciprocal along with certain types of ferrite phase shifters. There are basic types of phase shifters: diode phase shifters, nonreciprocal ferrite phase shifters, and reciprocal (dual-mode) ferrite phase shifters. Each has its properties, and the choice for using is highly dependent on the radar requirements. No type of phase shifter is universal enough to meet the requirements of all applications.

A signal of wavelength λ passing through a line of length l at a velocity v yields phase shift

$$\varphi = 2\pi f t = 2\pi f \frac{l}{v} = 2\pi f \frac{l}{c/\sqrt{\varepsilon_r \mu_r}} \qquad (9.7-17)$$

where

$c =$ velocity of light,
$f =$ frequency,

$\varepsilon_r =$ permittivity,

$\mu_r =$ permeability.

Therefore the equation (9.7-17) indicates that changing f, l, μ_r, ε_r, v can alter phase shift.

In the past, electronic phase shifters were analog. Their phase shift can be continuously adjusted. At present, Digital phase shifters are widely used. For example, an N-bit phase shifter covers 360° of phase change in 2^N steps. A four-bit digital phase shifter is a cascade of 22.5°, 45°, 90° and 180° phase shift units as shown in Figure 9.7-5. The least phase shift is $\Delta \varphi = 360/2^4 = 22.5°$. Each phase shift unit is controlled by binary values (i.e. "1" or "0"). "1" corresponds to phase shift, while "0" corresponds to no phase shift. For example, if a control signal is 1010, then the phase shift of the phase shifter is

$$\varphi = 1 \times 180° + 0 \times 90° + 1 \times 45° + 0 \times 22.5° = 225°$$

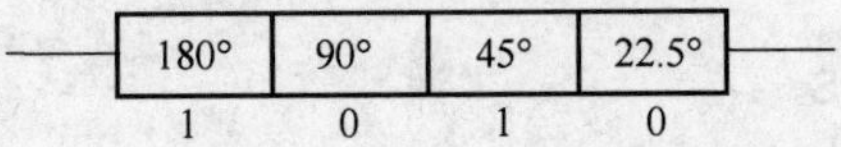

180°	90°	45°	22.5°
1	0	1	0

Figure 9.7-5 Four-bit digital phase shifter

9.7.4 Angle Measurement of Phased Array Radar

Angle measurement of phased array radar is also based on phase-comparison monopulse technique. In phased array radar, two methods for angle measurement are used.

1. Phase comparison

The principle of the phase comparison angle measurement is shown in Figure 9.7-6.

A line array can be divided into two sub-line-arrays. The distance between their centers is D $\left(D = \frac{1}{2}nd, d =\right.$ element interval$\left.\right)$. The phase difference $\Delta \varphi$ between the two signals, which the two sub-line-arrays receive from θ direction, is

$$\Delta \varphi = \frac{2\pi}{\lambda} D \sin \theta \qquad (9.7-18)$$

Two receivers provide two orthogonal components, I and Q. They are converted by A/D converter for sending them to the digital signal processor. The phases of the signals which two receivers produce are

$$\left. \begin{array}{l} \varphi_1 = \arctan(Q_1/I_1) \\ \varphi_2 = \arctan(Q_2/I_2) \end{array} \right\} \qquad (9.7-19)$$

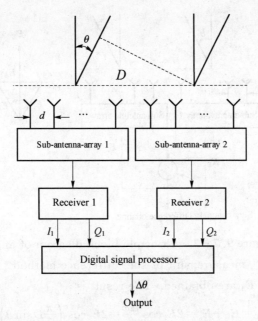

Figure 9.7 – 6 Principle block diagram of phase comparison angle measurement

The phase difference is

$$\Delta\varphi = \varphi_1 - \varphi_2 \qquad (9.7-20)$$

In order to measure the elevation and azimuth angles, a planar array is needed. It may be divided into 4 parts. Thus, 4 receivers are needed.

2. Sum-difference method in phased array radar

Figure 9.7 – 7 shows a principle block diagram of angle measurement of sum-difference method.

In the distance, the effect of the range difference on the amplitude may be neglected, and the amplitudes are same. When the target deviates from the boresight, the signal which two antenna arrays receive has the phase difference due to the wave-way difference. The phase difference is

$$\Delta\varphi = \frac{2\pi}{\lambda} D \sin\theta \qquad (9.7-21)$$

The amplitudes of echoes which the two sub-array-antennas receive are equal, i.e. $E_1 = E_2$, and their phase difference is $\Delta\varphi$. The hybrid-ring junction may produce sum signal and difference signal. Referring to the right-hand portion of Figure 9.7 – 7, the sum signal E_Σ

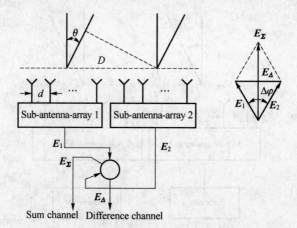

Figure 9.7-7 Principle block diagram of angle measurement of sum-difference method

and the difference signal E_Δ are obtained. The result is

$$\left.\begin{aligned} E_\Sigma &= E_1 + E_2 = 2E_1 \cos \frac{\Delta\varphi}{2} = 2E_1 \cos\left(\frac{\pi}{\lambda} D \sin\theta\right) \\ E_\Delta &= E_1 - E_2 = 2E_1 \sin \frac{\Delta\varphi}{2} = 2E_1 \sin\left(\frac{\pi}{\lambda} D \sin\theta\right) \end{aligned}\right\} \quad (9.7-22)$$

Assuming that the beam direction angle is θ_0, and the target direction angle is $\theta = \theta_0 + \Delta\theta$, where $\Delta\theta$ is less than the half power beamwidth, The equation (9.7-21) becomes

$$\Delta\varphi = \frac{2\pi}{\lambda} D \sin(\theta_0 + \Delta\theta) = \Delta\varphi_0 + \delta\varphi \quad (9.7-23)$$

where $\Delta\varphi_0$ is the phase difference when the target is located in the beam direction, which is estimated from the beam control parameter that the beam controller supplies. Thus,

$$\Delta\varphi_0 \approx \frac{2\pi}{\lambda} D \sin\theta_0 \quad (9.7-24)$$

In view of small $\Delta\theta$, $\sin\Delta\theta \approx \Delta\theta$, the equation (9.7-23) minus the equation (9.7-24) yields

$$\delta\varphi = \Delta\varphi - \Delta\varphi_0 = \frac{2\pi}{\lambda} D \cos\theta_0 (\Delta\theta) \quad (9.7-25)$$

In point of $\Delta\varphi = \Delta\varphi_0 + \delta\varphi$, where $\Delta\varphi_0$ is compensated by way of the phase shifters, actually, the phase difference of both signals which are fed to hybrid-ring junction is $\delta\varphi$.

Substitute equation (9.7-25) into equation (9.7-22) to yield

$$\left.\begin{aligned} |E_\Sigma| &= 2|E_1| \cos\left(\frac{\pi}{\lambda} D \cos\theta_0 \times \Delta\theta\right) \\ |E_\Delta| &= 2|E_1| \sin\left(\frac{\pi}{\lambda} D \cos\theta_0 \times \Delta\theta\right) \end{aligned}\right\} \quad (9.7-26)$$

Figure 9.7-7 indicates that there is $\pi/2$ phase shift between E_Δ and E_Σ. When the target locates at two sides of the beam direction, the direction of E_Δ alters 180° as is the same to the phase-comparison monopulse.

$|E_\Delta|/|E_\Sigma|$ is normalized to become

$$K(\theta) = \frac{|E_\Delta|}{|E_\Sigma|} = \tan\left(\frac{\pi}{\lambda} D \cos\theta_0 \times \Delta\theta\right) \qquad (9.7-27)$$

Therefore, the deviation angle $\Delta\theta$ from the beam direction θ_0 is

$$\Delta\theta = \frac{\lambda}{\pi D \cos\theta_0} \arctan[K(\theta)] \qquad (9.7-28)$$

If $\Delta\theta$ is small, moreover $\Delta\theta \ll \Delta\theta_{0.5}/2$, the phased array tracks the target steadily, and the target holds at the neighborhood of the maximum beam value θ_0, then equation (9.7-27) becomes

$$K(\theta) = \frac{\pi}{\lambda} D \cos\theta_0 \times \Delta\theta \qquad (9.7-29)$$

Thus,

$$\Delta\theta = \frac{\lambda}{\pi D \cos\theta_0} K(\theta) \qquad (9.7-30)$$

9.7.5 Range and Angle Tracking System of Phased Array Radar

Figure 9.7-8 illustrates a range and angle track system of phased array radar. The echo which the phased array antenna receives is divided into three way signals, which are sum Σ signal, azimuth difference ΔA signal, and elevation difference ΔE signal. They are fed to the monopulse receiver. After they are amplified at high frequency, mixed, amplified at intermediate frequency, and detected phase orthogonally, the six way signals $\Sigma_I, \Sigma_Q, \Delta A_I, \Delta A_Q, \Delta E_I$, and ΔE_Q are produced. Making use of Σ_I and Σ_Q signals, and the early and late gates, the range error processor makes conversion and manipulation, and then yields normalized range error signal ΔR, which are extrapolated in digital filter to yield the predictive value R_p of the target range. The R_p is used to control the digital time modulator and track gate generator so as to form a closed range track loop. As a result, the range tracking for the target will hold. The digital time modulator also triggers the range gate generator to produce the range gates, which strobe the mono-pulse receiver and error signal processors in range for multi-channels control assignment and selection in multiple-target condition. Making use of $\Sigma_I, \Sigma_Q, \Delta A_I, \Delta A_Q, \Delta E_I, \Delta E_Q$, and the range gate, the angle error processor makes conversion and manipulation to produce normalized azimuth angle error signal $|\Delta A|/|\Sigma|$ and elevation angle error $|\Delta E|/|\Sigma|$, which are extrapolated in digital filter to generate the predictive values A_p and E_p of the azimuth and elevation angles. In light of the predictive values A_p and E_p, the beam control computer computes phase shift of each phase shifter of the phased array to control phase of each antenna element by the interface driver for correcting the beam direction. Its purpose is to make the phased array antenna beam accurately aim at the target.

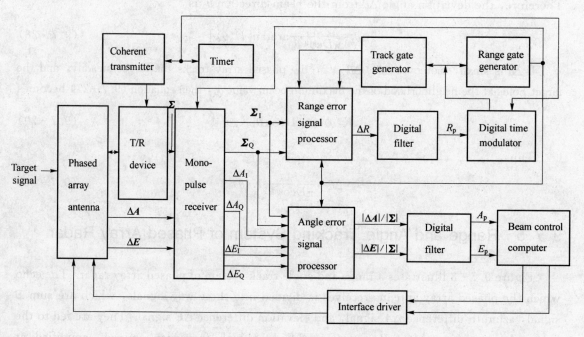

Figure 9.7-8 Range and angle track system of phased array radar

9.7.6 Brief Introduction to Multi-function Phased Array Radar

The multi-function phased array radar includes search, track, weapon control, missile guidance, target recognition, and perhaps others. American Patriot missile weapon system, Aegis missile weapon system, and Russian C—300 missile weapon system employ the multi-function phased array radar. It has the following advantages.

① Rapid reaction. The time from search to track is short. For example, for Patriot missile weapon system, the reaction time is shortened to 10 seconds.

② Entire airspace work. For example, in Aegis ship-carrier missile weapon system, the 4 array planes can cover 360°. For another example, in C—300 weapon system the phased array is installed to a platform, which rotates 360° in azimuth.

③ Multiple-target manipulation. Because using electronic scan, the beam direction changes flexibly. The track-while-search manner or track and search may be chosen. Patriot missile weapon system can scout 100 targets, track 9 targets, and guide 9 missiles.

④ Strong anti-jamming. It uses time region processing, frequency region processing, and space region processing for anti-jamming.

9.8 Command Guidance System Design

Three-point guidance is widely used in command guidance systems. A design example of three-point guidance is given below.

The block diagram of three-point guidance can be drawn as shown in Figure 9.8 - 1.

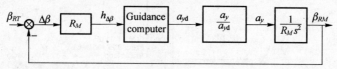

Figure 9.8 - 1 Block diagram for three-point guidance

The a_y/a_{yd} transfer function was given in equation (8.5 - 6), which is repeated here for convenience:

$$\frac{a_y}{a_{yd}} = \frac{113\,400(-0.36s^2 - 1.08s + 1\,320)}{(s+21.4)(s^2+18.94s+254.8)(s^2+145.6s+27\,450)} \quad (9.8-1)$$

A double integral element is required for transforming an acceleration into a distance. A lead compensator is required for stabilization.

Use RLTOOL of MATLAB (SISO Design Tool) to design guidance loop. MATLAB program 9.8 - 1 gives the SISO Design Tool. The SISO Design Tool is shown in Figure 9.8 - 2.

MATLAB program 9.8 - 1

```
>> num1=[113400];den1=[1 21.4];G1=tf(num1,den1);
>> num2=[-0.36 -1.08 1320];den2=conv([1 18.94 254.8],[1 145.6 27450]);
>> G2=tf(num2,den2);G3=G1*G2;
>> num4=[1];den4=[1 0 0];G4=tf(num4,den4);
>> G5=G3*G4;
>> rltool(G5);
```

The result of simulation test indicates that adding a compensator $G_{comp} = \dfrac{2(1+2s)}{(1+0.05s)}$ will produce satisfactory frequency domain indexes. Figure 9.8-2 shows the gain margin and the phase margin as follows:

$$Gm = 7.01 \text{ dB (at 8.27 rad/s)}, \quad Pm = 42.2° \text{ (at 3.97 rad/s)}$$

The gain margin and phase margin are healthy.

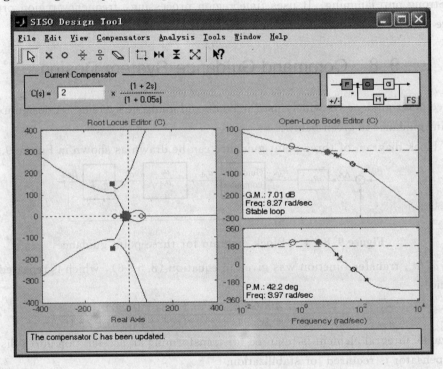

Figure 9.8-2 SISO Design Tool of the guidance loop of three-point method

MATLAB program 9.8-2 generates the closed system step response. The resulting step response is shown in Figure 9.8-3.

```
MATLAB program 9.8-2
>> n=[4 2];d=[0.05 1];G6=tf(n,d);
>> G7=G5*G6;
>> G=feedback(G7,1,-1);
>> step(G,5);
```

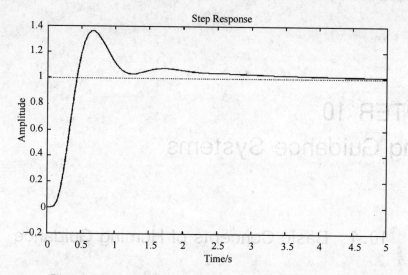

Figure 9.8-3   Step-response of the guidance loop

References

[1] Garnell P. Guided weapon Control Systems. Oxford:Pergamon Press,1980.
[2] John H Blakelock. Automatic Control of Aircraft and Missiles. 2nd ed. New York:Wiley,1991.
[3] Skolnik M I. Introduction to Radar System. 3rd ed. New York:MCGraw-Hill Book Company, 2001.
[4] Skolnik M I. Radar Handbook. 2nd ed. New York:MCGraw-Hill Book Company, 1990.
[5] David K Barton. Radar System Analysis. Englewood Cliffs,New Jersey:Prentice-Hall, Inc. ,1964.
[6] 陈佳实. 导弹制导和控制系统的分析与设计. 北京:宇航出版社, 1989.
[7] 秦忠宇. 防空导弹制导雷达跟踪系统与显示与控制. 北京:宇航出版社,1995.

CHAPTER 10
Homing Guidance Systems

10.1 Basic Concepts of Homing Guidance

10.1.1 Components of Homing Guidance System

Homing guidance system depends on onboard homing head to receive a certain characteristic energy (microwave or optical) which a target radiates or reflects to extract tracking data, and to compute control commands in order that the missile is steered to fly to the target. A block diagram for a homing guidance system is indicated in Figure 6.7-1.

The homing guidance system is composed of the following parts:

① Homing head. It is a key device. According to radiating or reflecting signals of a target, it acquires, tracks the target, and yields guidance parameters.

② Device computing guidance command. In light of selected guidance law, it synthesizes the guidance parameters to produce guidance commands.

③ Stabilization loop of the missile. It consists of the autopilot and the airframe, generally including the damping loop and the acceleration control loop. Its main function is to stabilize the missile's attitude and to receive the guidance command.

④ Missile-target kinematics. It is composed of kinematical equations, which describe the relative motion relation between the missile and the target.

Compared with the remote control guidance, the homing guidance has a higher accuracy, but its operating range is shorter, and the onboard device is complex.

10.1.2 Classification of Homing Guidance

1. Classification based on detecting frequency spectrum

According to detecting frequency spectrum, there are microwave, optical, millimeter wave, laser, and multi-mode guidance.

(1) Microwave homing guidance

It can work in all weather, mainly operate in centimeter band. Its operating range is longer than optical homing guidance, but its space resolution is lower than that of optical homing guidance.

(2) Optical homing guidance

It operates in infrared or visible light band. As its wavelength is shorter, its space resolution is higher. However, it is susceptible to weather, and when optical signals pass through cloud, fog, and rain, there are greater attenuations. Visible light guidance can not operate at night.

(3) Millimeter wave homing guidance

As its frequency spectrum locates between the infrared and microwave bands, its performance also lies between infrared guidance and microwave guidance.

(4) Laser homing guidance

Because the laser has good directivity and narrow wave beam, it has high guidance accuracy. Besides, since laser spectrum brightness is high, and homochromatism is good, it has stronger anti-jamming ability and recognizing target ability. It operates day and night. But, compared with microwave and millimeter band, its attenuation in atmosphere is greater, and its ability to penetrate through smog is weaker. Generally, it is used for short range missiles, guidance bombs, and cannonballs.

(5) Multi-mode guidance

It employs two or more frequency bands, such as radio and infrared frequencies, for guidance. Since it can absorb each frequency band advantages, it has strong anti-jamming and anti-stealth abilities.

2. Classification based on detecting energy source

In light of detecting energy source, homing guidance is classified into active homing guidance, semi-active homing guidance and passive homing guidance.

(1) Active homing guidance

An active homing guidance is a homing guidance in which both the transmitter and the receiver are carried on board. The active homing guidance system has an onboard active homing head, which can illuminate the target and receive the reflected energy from the target. According to reflected energy, the active homing head can detect the presence of the target, track the target, and provide guidance information. It is used to generate guidance commands. They direct the missile flight to intercept the target. Once the homing head acquires the target and shifts to track the target, the missile completely independently implements guidance, but no requirement for external guidance intelligence. Thus, active guided missiles have the advantage of launch-and-leave. The onboard guidance device is complicated, and the missiles are burdened with additional weight. As limited by illuminating source on board, the range is shorter. The energy used to illuminate the target is commonly in the form of radio. An example of an active homing missile system is the European Meteor active radar-guided AAM (air-to-air missile).

(2) Semi-active homing guidance

In semi-active homing guidance, the external guidance station provides the energy to illuminate the target. The illuminating source may be ground-based, ship-borne, or airborne. The principal difference between the active homing and the semi-active homing is that the semi-active homing is dependent on external sources. The guidance process of the semi-active homing is similar to that of the active homing. Since the semi-active homing guidance requires the external radar to continuously illuminate the target at all times during the flight of the missile, the illuminating radar is prone to being attacked. As the off-board antenna aperture is not limited by the missile space, the operating range is greater. The semi-active guidance is widely applied to ground-to-air missiles, ship-to-air missiles, air-to-air missiles, air-to-ground missiles, and anti-tank missiles. An example of a semi-active homing missile system is the supersonic Sparrow III (model AIM—7F).

(3) Passive homing guidance

A passive homing guidance is one in which the onboard receiver makes use of energy emanating from the target. As the target supplies guidance energy, the illuminator is not required. The guidance process of the passive homing is similar to that of the other two homing guidance. The energy emanating from the target may be in the forms of heat, light, or radio waves, which correspond to infrared homing guidance, visible light homing guidance, and anti-radiation homing guidance respectively. Passive guided missiles have the advantage of concealment, but the operating range is shorter. The Sidewinder missile is an example of a passive infrared homing guided missile.

10.1.3 Rate of Change of Line-of-sight

In homing guidance, the proportional navigation is commonly employed. In other words, the guidance law is formed based on the LOS (line-of-sight) rate $\dot{q}$. Assuming that the proportional navigation is used, Figure 10.1-1 shows the change process of the LOS rate $\dot{q}$, which is approximately divided into three phases: eliminating initial error phase, accurately tracking phase, and divergent phase.

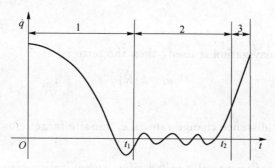

Figure 10.1-1 $\dot{q}$ change process

The first phase is to eliminate initial error. At the beginning of the guidance, there is initial LOS rate, which rests with initially sighting error. With the operation of the homing head, the error gradually decreases. The adjusting time lies on initial LOS rate, available load factor (normal acceleration) of the missile, and inertia of guidance loop. When the $\dot{q}$ is less than a given value, the first phase is over, and then shifts to the second phase.

The second phase is the accurate tracking phase. As there are the inertia and disturbance in guidance system, $\dot{q}$ fluctuates around zero value. The fluctuating amplitude determines dynamic error of the guidance.

The third phase is divergent phase. When the missile-target distance gradually becomes shorter, $\dot{q}$ tends to diverge. $\dot{q}$ is

$$\dot{q} = \frac{\Delta V_v}{r} \qquad (10.1-1)$$

where

ΔV_v = projection of the missile-target relative velocity onto the direction perpendicular to the LOS,

r = missile-target distance.

CHAPTER 10   Homing Guidance Systems

From equation (10.1-1), it can be seen that if $r \to 0$, $\dot{q} \to \infty$.

10.1.4 Guidance Command of Homing Guidance System

Similar to the command guidance system, the homing guidance includes two parts: guidance law component, compensation component. In addition, it yields the fuse control command opening safety in impact phase. The guidance commands are composed of pitch control command and lateral control command. As the two commands are similar, the pitch command will be discussed as follows.

1. Principal term

If the proportional navigation is used, then the term is

$$u_k = k|\dot{R}|\dot{q} \qquad (10.1-2)$$

where

$\dot{R}$ = missile-target distance change rate, i.e. missile-target closing velocity,

$\dot{q}$ = LOS rate,

k = proportional coefficient, also called effective navigation ratio, $k \geqslant 2$, generally taken on 3~6. For the ground-to-air missile, if the span of air area is great, the variable coefficient may be used. At long distance (initial guidance and mid guidance phases), it is taken small value, and trajectory is straight such that the maneuver of missile is decreased to save the energy in favor of increasing range or increasing maneuverability of terminal phase. At close quarters (terminal phase), it should be increased to rapidly eliminate the error and speedily hit the target.

Sometimes it is difficult to pick up $\dot{R}$ (for example, infrared point source homing head). In this case, use $u_k = k'\dot{q}$ in stead of $u_k = k|\dot{R}|\dot{q}$.

In actual homing guidance system, the LOS $\dot{q}$ is replaced by the error of LOS angle u_ϵ, which contains measurement noise, so the filter is used. In addition, the compensator should be designed for the sake of stabilization of the guidance loop.

2. Compensation term of tangent acceleration

When the tangent acceleration direction is inconsistent with the line of sight, additional angular speed of line of sight occurs and arouses the trajectory to be curved. Using the signal with equal value in magnitude and opposite sign may cancel completely the effect. Actually, we can not cancel completely, only approximately compensate. Generally, the following term is taken to approximately compensate:

$$u_a = k_a a_t \varphi \qquad (10.1-3)$$

where

$a_t =$ tangent acceleration,
$\varphi =$ angle between LOS and longitudinal axis of missile,
$k_a =$ proportional coefficient.

3. Compensation term of gravity

The gravity of the missile gives disturbance to the pitch channel of the guidance loop, so the gravity effect should be compensated in this channel. In general, the following compensation is taken:

$$u_g = k_g g \cos \gamma \qquad (10.1-4)$$

where

$\gamma =$ flight path angle,
$k_g =$ proportional coefficient.

In actual homing guidance system, the magnitudes of the guidance commands need to be limited so as to avoid exceeding available normal acceleration of the missile.

In practice, there may be preset commands for the homing head pointing and for the missile flight path as well.

10.2 Homing Heads

10.2.1 Introduction to Homing Heads

A homing head is sometimes called a seeker or homing eye. By receiving emanating signal or reflecting signal of a target, a homing head tracks the target, measures angular speed of line of sight, and produces guidance information.

1. Main functions of homing head

Main functions of homing head include:
① Receive pre-arranging parameter from the fire control system,
② Acquire the target rapidly,
③ Perform automatic target tracking,

④ Decrease disturbance of airframe, stabilize direction of antenna for radar seeker, or stabilize direction of light axis for optical seeker,

⑤ Measure relative motion parameter of missile-target,

⑥ Produce guidance information,

⑦ Transmit relational parameter or command to the fuse.

2. Classification of homing heads

In light of energy source, there are active homing heads, semi-active homing heads, and passive homing heads.

In terms of source characteristics, there are radar homing heads, television homing heads, infrared homing heads, and laser homing heads.

3. Components of a homing head

In essence, the seeker is an onboard tracker. An infrared (IR) seeker typically consists of the following components: ① a gimbaled platform that contains the optical components for collecting and focusing the target radiation, ② an IR sensor that converts the incident radiation into electrical signals, ③ the electronics for processing the sensor output signals and converting them into guidance commands, ④ a servo and stabilization system to control the position of the tracking platform, ⑤ IR cooling system, and ⑥ a protective covering, i. e. the dome (also known as irdome).

Figure 10.2 – 1 illustrates an optical system of an IR seeker. It includes an irdome, a primary mirror, a secondary mirror, a correcting mirror, an optical filter, and a field mirror. The infrared radiation incident upon the seeker dome passes through the dome and strikes the

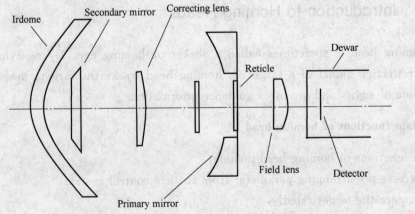

Figure 10.2 – 1   Optical system of an IR seeker

primary mirror, which in turn redirects the incident radiation to the secondary mirror. The mirror reflects the radiation to the correcting lens, which focuses the radiation on the reticle. The reticle periodically modulates the incoming radiation so as to extract and track a target. Finally, the field lens condenses the radiation to the IR detector. The optical filter filters the beyond operating frequency radiation, and permits the predetermined spectrum radiation to pass through.

Radar seekers comprise the components similar to those of optical seekers. For example, Figure 10.2 – 2 shows a simplified block diagram of a conical-scan semi-active seeker. It includes a forward-wave antenna, a forward-wave receiver, an echo antenna, an antenna receiver, a speed tracking circuit, an angle error generator, a conical-scan motor, and servos of antenna. Besides, there is a radome at the foreside of the homing head, which protects it from the effect of environment outside missile.

The forward-wave signal and the echo signal are compared to produce a target doppler frequency. The speed tracking provides closing velocity information. The angle error generator provides LOS angle rate information. The command generator yields proportional navigation commands.

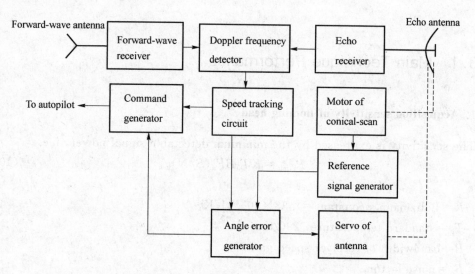

Figure 10.2 – 2 Block diagram of conical-scan semi-active seeker

10.2.2 Radar Homing Heads

The radar homing head receives the electromagnetic wave that the target reflects or

CHAPTER 10 Homing Guidance Systems

emanates, detects relative position of missile-target, tracks the target, picks up guidance intelligence, and produces guidance commands to autopilot of missile. Its main functions include:

① Acquire and track target,
② Extract guidance intelligence and produce guidance commands,
③ Decrease the effect of the airframe disturbance on the antenna direction.

In light of the electromagnetic wave spectrum, there are microwave homing head and millimeter homing head.

In terms of mode abstracting angle signal, there are scanning seekers, monopulse seekers, and phased array seekers.

According to operating wave, the radar homing heads are classified into continuous wave and pulse doppler homing heads.

In view of energy source, there are active radar homing head, semi-active radar homing head, and passive radar homing head.

10.3 Semi-active Radar Homing Heads

10.3.1 Main Technique Performance

1. Acquisition sensitivity of homing head

The sensitivity is expressed by the minimum detectable signal power, i.e.
$$P_{r,\min} = KT_0 BF_n (S/N)_{\min} \tag{10.3-1}$$

where

$K =$ Boltzmann's constant $= 1.38 \times 10^{-23}$ J/K,
$T_0 =$ standard temperature, 290 K,
$B =$ bandwidth of receiver speed gate,
$F_n =$ noise factor,
$(S/N)_{\min} =$ minimum output signal-to-noise ratio, generally taken on the values of 3～5 dB.

In order to increase the sensitivity, i.e. decrease minimum detectable signal power, the key is to decrease noise factor. Generally, people use low noise high frequency amplifier, match filter in intermediate frequency, and narrow speed-gate width.

2. Beam width of seeker antenna

The beam width of seeker antenna is expressed by half power beam width $\theta_{0.5}$. $\theta_{0.5}$ is

$$\theta_{0.5} = K \frac{\lambda}{D} \times 57.3 \qquad (10.3-2)$$

where

K = proportional coefficient, which has relationship with the antenna efficiency, generally taken on the values of $1 \sim 1.3$,

λ = operating wavelength,

D = antenna aperture.

Decreasing the operating wavelength or increasing the antenna aperture may reduce the beam width $\theta_{0.5}$, and then the space resolution increases so as to create advantageous condition for more accuracy. The antenna aperture is limited by the missile space, so higher operating frequency is generally used, such as millimeter wave seekers.

3. Operating band

For seeker, optional band includes: C-band ($4 \sim 8$ GHz), X-band ($8 \sim 12.5$ GHz), Ku-band ($12.5 \sim 18$ GHz), Ka-band ($26.5 \sim 40$ GHz), millimeter band (8.5 millimeter, 3.2 millimeter, 1.4 millimeter). When selecting operating band, we mainly consider resolution, transmission loss, and transmitter power. The higher the operating frequency, the higher the space resolution will be. Therefore the developments of radar seekers trend to high frequency.

4. Range

For the semi-active seeker, the equation of the range is

$$(R_T R_R)_{min} = \left[\frac{P_t G_t G_r \lambda^2 \sigma_T F_T^2 F_R^2}{(4\pi)^3 P_{r,min} L_T L_R} \right]^{\frac{1}{2}} \qquad (10.3-3)$$

where

R_T = transmitter-to-target range,

R_R = receiver-to-target range,

P_t = transmitter power,

G_t = transmit antenna power gain,

G_r = receive antenna power gain,

λ = operating wavelength,

σ_T = target radar cross section (RCS),

$P_{r,\min}$ = seeker sensitivity,

L_T = transmit system loss,

L_R = receive system loss,

F_T = pattern propagation factor of transmitter-to-target path,

F_R = pattern propagation factor of target-to-receiver path.

The range of the forward-wave is

$$R_0 = \left[\frac{P_t G_D G_W \lambda^2}{(4\pi)^2 P_{r,\min} L_T}\right]^{\frac{1}{2}} \quad (10.3-4)$$

where

P_t = illuminating power of forward wave,

G_D = antenna gain of illuminating radar,

G_W = rear forward-wave receiving antenna gain,

$P_{r,\min}$ = sensitivity of rear receiver,

L_T = transmit system loss,

R_0 = range from illuminating radar to missile.

When a missile flies to a target, the missile-target distance becomes shorter, and the echo signal becomes stronger. If the echo intensity exceeds dynamic range of the receiver of the seeker, the receiver saturates and the seeker is invisible to the target. The phenomena is called dazzling effect of target. The invisible range is

$$R_{LS} \leqslant \frac{1}{R_T}\left[\frac{P_t G_t G_r \lambda^2 \sigma_T}{(4\pi)^3 P_L}\right]^{\frac{1}{2}} \quad (10.3-5)$$

where P_L is the signal power, which the seeker receives when being invisible, other symbols are defined ibid.

5. Technique requirement of servo stabilization

(1) Uncoupling coefficient

It represents ability that the seeker eliminates the airframe disturbance to the line of sight. As the seeker is installed on board, the missile attitude angle rate $\Delta\dot{\theta}$ influences LOS rate $\dot{q}$ to produce the measurement error. Therefore, in seeker there is a stabilization loop or a stabilization platform. When the angle rate $\Delta\dot{\theta}$ disturbance occurs, the rate gyro in seeker can sense the antenna angular speed and provide feedback to the servo of the antenna to rotate in opposite direction. As a result, this stabilizes the antenna pointing. Input $\Delta\dot{\theta}$ yields the LOS rate $\Delta \dot{q}$. The ratio of $\Delta \dot{q}$ to $\Delta \dot{\theta}$ is the uncoupling coefficient k. Actually, the seeker is a dynamic element, so its output has relationship with the frequency. When the frequency increases, uncoupling ability decreases. If ignoring the frequency effect, the uncoupling coefficient is

$$k = \Delta \dot{q} / \Delta \dot{\theta} = 1/(1+K) \approx 1/K \qquad (10.3-6)$$

where K is the open loop gain of the antenna servo.

In general, if the frequency is less than 3 Hz, the uncoupling coefficient should be less than 0.05.

(2) Maximum tracking coverage

The maximum tracking coverage directly influences lethal area, and is generally $\pm 45° \sim \pm 60°$.

(3) Maximum tracking angular speed

It generally takes the value of $10(°)/s \sim 40(°)/s$.

(4) Minimum tracking angular speed

It depends on the low speed performance of the servo. It is generally $0.01(°)/s \sim 0.05(°)/s$.

6. Technique requirements for radomes

For the antenna pointing, the radome plays a negative role, and leads to the pointing error, so there are basic technique requirements for radomes as follows.

(1) Power transmission coefficient

Generally requires that it is greater than 85%.

(2) Error of line of sight

Generally requires that it has smaller error of line-of-sight.

(3) Slope of error of line of sight

It is defined as a derivative of aberration angle (the error of LOS) with respect to look angle. Detailed definition is described in subsection 10.3.2. As the slope may be positive or negative, generally requires its absolute value is less than 0.05.

10.3.2 Work Principle of a Semi-active Continuous Wave Seeker

The concept of semi-active homing guidance is indicated in Figure 10.3-1. In guidance process, an illuminator illuminates a target, maintains the target within its radar beam, and

CHAPTER 10 Homing Guidance Systems

provides a reference to the missile. The missile receives the target-reflected echo from its front antenna and a sample of the directly received illumination (often through sidelobes of the illuminator antenna) by its rear antenna. The front and rear signals are coherently detected against each other, resulting in a spectrum which contains the doppler-shifted target signal at a frequency roughly proportional to closing velocity. A narrowband frequency tracker searches the spectrum, locks onto the target echo, and extracts guidance information from it. The use of CW (continuous wave) radar enables the system to discriminate against clutter on the basis of doppler frequency and thus allows low-altitude operation.

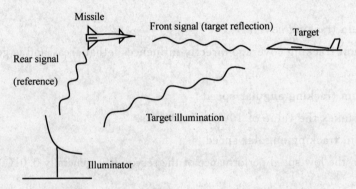

Figure 10.3 – 1 Semi-active homing guidance

1. Frequency relation

Figure 10.3 – 2 shows the geometrical relation among the illuminating radar, the missile, and the target. The doppler shift is

$$f_{\text{dop}} = (f_0/c)V_R = V_R/\lambda$$

where

$f_0 =$ transmitting frequency,

$V_R =$ the component of velocity along the line of sight from the source to the observer—either a receiver or a reflector,

$c =$ velocity of light,

$\lambda =$ wavelength.

The forward-wave signal frequency which the rear antenna of the missile receives is

$$f_R = f_0 + f_{dr} \qquad (10.3-7)$$

where

$f_0 =$ the transmitting frequency of the illuminating radar,

f_{dr} = the doppler frequency shift which results from the relative motion between the missile and the illuminating radar.

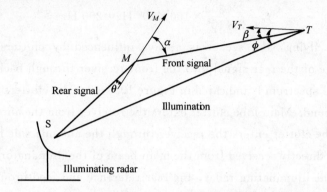

Figure 10.3-2 Geometrical relation among the illuminating radar, the missile, and the target

From Figure 10.3-2,

$$f_{dr} = -\frac{V_M \cos\theta}{\lambda} \qquad (10.3-8)$$

where λ is the wavelength of illuminating wave.

The echo signal frequency which the fore antenna of the missile receives is

$$f_F = f_0 + f_{df} \qquad (10.3-9)$$

where f_{df} is the doppler frequency shift which results from the relative motions between the target and the illuminator, and between the missile and the target. The doppler frequency shift f_{df} which the fore antenna receives is

$$f_{df} = \frac{V_T}{\lambda}\cos\phi + \frac{V_T}{\lambda}\cos\beta + \frac{V_M}{\lambda}\cos\alpha \qquad (10.3-10)$$

When the front signal is coherently detected (mixed) against the rear signal, the resulting spectrum is the difference of the two, i.e.

$$f_d = f_{df} - f_{dr} = \frac{V_T}{\lambda}\cos\phi + \frac{V_T}{\lambda}\cos\beta + \frac{V_M}{\lambda}\cos\alpha + \frac{V_M}{\lambda}\cos\theta \qquad (10.3-11)$$

For the head-on case, ϕ, β, and θ are considered to be small, and approximately zero, so

$$f_d = \frac{2\times(V_T+V_M)}{\lambda} = \frac{2V_C}{\lambda} \qquad (10.3-12)$$

where $V_C = V_T + V_M$ is the relative speed of the missile-target. The equation indicates that the doppler frequency shift is proportional to the relative speed, and inversely proportional to the transmitted wavelength. For example, if a closing velocity (relative speed) $V_C = 300$ m/s (1 080 km/h), and a transmitted frequency $f = 100$ MHz, the doppler frequency is

$$f_d = \frac{2 \times 300}{3 \times 10^8} \times 100 \times 10^6 \text{ Hz} = 200 \text{ Hz}$$

If a missile is flying, the seeker is always influenced by clutters and feedthrough (spillover or leakage of the rear signal into the front receiver through backlobes of the front antenna). A typical spectrum is indicated in Figure 10.3-3. The clutters come from ground objects and background. Main-lobe clutter enters the receiver from the antenna main-beam of the seeker. Side lobe clutter enters the receiver through the antenna side lobe of the seeker. The feedthrough is directly received from the main-beam of the illuminator. Since the missile moves away from the illuminating radar, the rear reference and feedthrough frequency shift down a doppler frequency for V_M/λ relative to the transmitting frequency. The target echo signal frequency shifts upward for $(2V_T + V_M)/\lambda$. The maximum clutter occurs at $f_0 + V_M/\lambda$ position. On the basis of the rear reference, the feedthrough occurs at zero frequency, the clutter maximum occurs at $2V_M/\lambda$, and the target peak occurs at $2(V_T + V_M)/\lambda$, or $2V_C/\lambda$. For instance, for the X-band case, let $V_M = 2\ 000$ ft/s and $V_T = 500$ ft/s in level flight. The main-lobe clutter (MLC) will then be at roughly 40 kHz and the target at 50 kHz.

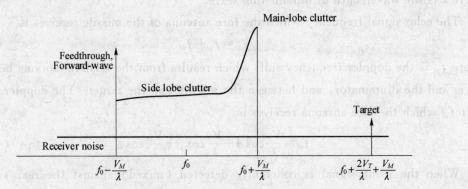

Figure 10.3-3 Clutter spectrum

Although the clutter and feedthrough are greater than the target signal, the target signal can be separated from the large interfering signals by using a narrowband filter with center frequency at $2V_C/\lambda$, because the target signal has the doppler frequency shift. When the target speed reduces, the target spectrum can approach the main lobe area, and it is difficult to detect the target.

2. Basic semi-active seeker

Figure 10.3 – 4 illustrates an early continuous wave semi-active seeker, which was used in Hawk missile. The CW (continuous wave) seeker includes a rear receiver, a fore receiver, a signal processor (speed gate), and a tracking loop to control the gimbaled front antenna.

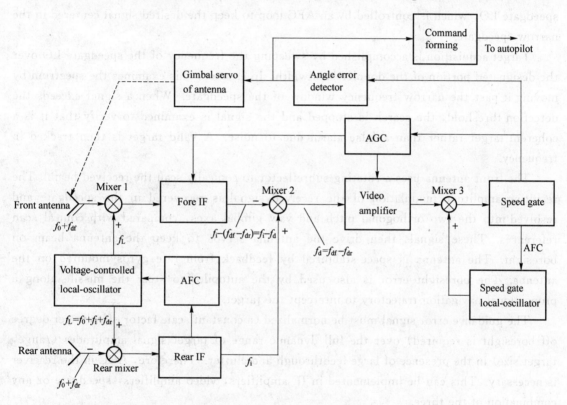

AFC—Automatic frequency control; AGC— Automatic gain control; IF—Intermediate frequency

Figure 10.3 – 4 Early continuous wave semi-active seeker

The purpose of the rear receiver is to provide a coherent reference for detection of the front (target) signal. It is an automatic frequency control loop, which consists of a rear mixer, a rear intermediate frequency amplifier, a voltage-controlled local-oscillator, an automatic frequency control circuit (containing a discriminator, a low-pass filter). When the loop is locked, the voltage-controlled local-oscillator frequency is tuned at $f_L = f_0 + f_i + f_{dr}$, where f_0 is illuminating frequency, f_i is intermediate frequency, and f_{dr} is the doppler frequency received by rear antenna. The forward-wave frequency for the rear antenna to receive is $f_0 + f_{dr}$. The rear mixer output frequency f_i, passes through the rear IF amplifier to provide a reference f_i for the mixer 2.

The echo frequency $f_0 + f_{df}$, received by the front antenna, is mixed with f_L to yield $f_L - f_0 - f_{df} = f_i - (f_{df} - f_{dr}) = f_i - f_d$. The front and rear IF amplifier outputs are mixed in mixer 2 to yield the doppler frequency shift $f_d = f_i - (f_i - f_d)$.

The doppler signal is amplified by the video amplifier, and then mixed with the speedgate LO, which is controlled by an AFC loop to keep the desired signal centered in the narrow speedgate.

Target acquisition is accomplished by sweeping the frequency of the speedgate LO over the designated portion of the doppler bandwidth. In essence, this examines the spectrum by moving it past the narrow frequency window of the speedgate. When a signal exceeds the detection threshold, the search is stopped and the signal is examined to verify that it is a coherent target rather than a false alarm due to noise. A valid target is then tracked in frequency.

The front antenna uses a rotating subreflector to conically scan the received beam. The resulting amplitude modulation of the received signal is recovered in the speedgate and resolved into the two orthogonal pitch and yaw gimbal axes, compared with conical scan references. These signals then drive the antenna servos to keep the antenna beam on boresight. The antenna is space-stabilized by feedback from rate gyros mounted on the antenna. The boresight error is also used by the autopilot to steer the missile along a proportional navigation trajectory to intercept the target.

The guidance error signal must be normalized (a constant scale factor of volts per degree off boresight is required) over the full dynamic range of target signal amplitudes (range, target size) in the presence of large feedthrough and clutter. Therefore, AGC in the receiver is necessary. This can be implemented in IF amplifiers, video amplifiers, speedgate or any combination of the three.

The receiver consists of a wide IF, a somewhat narrower doppler amplifier, and finally a narrowband speedgate, like an ever narrowing funnel. For example, typical values of the fore IF amplifier, the doppler video amplifier, and the speedgate bandwidths are 1 MHz, or wider, 100 kHz or wider, and 500～2 kHz respectively. In the receiver, the target signal must compete with feedthrough, clutter, and jamming until the final stages. The dynamic-range requirements of the receiver and its AGC loops are dictated by these large undesired signals. Since the frequency of the feedthrough is the same as the rear signal, for a system in which the front and rear signals are mixed directly (baseband conversion), feedthrough would occur at DC (zero frequency), with the approaching and receding spectra folded around it. Folding the spectrum around feedthrough folds the receiver noise as well, resulting in 3 dB higher noise level (hence a 3 dB loss in sensitivity).

3. Inverse receiver

A simplified block diagram of an inverse receiver is shown in Figure 10.3 − 5. The inverse receiver is based on the fact that the funnel of the conventional receiver is inversed, with the narrow speedgate bandwidth being placed in the IF, after only a nominal amount of fixed preamplifier gain. It was made possible by the availability of highly selective crystal filters at IF frequencies and low-noise tunable microwave sources. As a result, unwanted signals are eliminated very early in the signal path, thus reducing dynamic range requirements and avoiding any possible source of distortion. One additional conversion is used in the receiver to avoid the problem of too much gain at one frequency. The scheme is employed by improved Hawk missile.

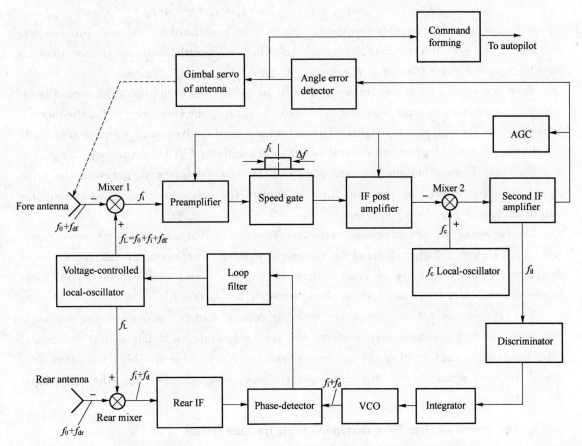

Figure 10.3 − 5 Block diagram of a inverse receiver seeker

Referring to Figure 10.3-5, the inverse receiver is composed of a forward-wave locked loop and a doppler frequency tracking loop, which is closed by the mixer 1 to form an automatic frequency control (AFC).

The forward-wave signal frequency received by the rear antenna is $f_0 + f_{dr}$. The echo signal frequency received by the fore antenna is $f_0 + f_{df}$. The relative doppler frequency shift is $f_d = f_{df} - f_{dr}$. When the loop tracks stably, the loop VCO frequency is $f_i + f_d$, and the forward-wave phase locked loop frequency f_L is $f_L = f_0 + f_i + f_{df}$. It is compared with the echo signal in mixer 1 to yield the echo difference frequency f_i, which is centered on the speed gate. When the doppler frequency f_{df} received by the front antenna changes, the voltage-controlled local-oscillator frequency follows the change to ensure the mixer 1 to yield the center frequency f_i, via the loop frequency automatic tracking.

The signal frequency provided by the rear mixer is

$$f_L - f_0 - f_{dr} = (f_0 + f_i + f_{df}) - (f_0 + f_{dr}) = f_i + f_d$$

where f_d is the target doppler frequency. Since it is modulated in the rear intermediate frequency, the rear intermediate frequency amplifier bandwidth must be greater than all possible target doppler frequency f_d range, i.e. $\Delta f_{ir} > f_{d,max} - f_{d,min} = \Delta f_d$.

Since too much gain at one frequency results in self-excited oscillation, the second local-oscillator or conversion local-oscillator is needed. Supposing the frequency is f_c, the mixer 2 provides $f_{i1} = f_c - f_i$ frequency signal. The second intermediate frequency signal is amplified, discriminated, and integrated to control speed local-oscillator (VCO) frequency. The VCO and the rear IF amplifier outputs are compared by phase-detector to generate the error signal, which is filtered by the loop filter to control the voltage-controlled local-oscillator output frequency f_L.

Inverse seeker is advantageous to clutter rejection and dynamical range requirement reduction. It sets a higher demand on electronic devices. For example, microwave local-oscillator frequency stability is greater than $10^{-8} \sim 10^{-9}$, and narrowband crystal filter bandwidth-to-center frequency ratio is about within $0.01 \sim 0.0001$.

The inverse receiver may cooperate with the conical scan or the mono-pulse for angle measurement. For a conical scan system, the scan sidebands must fall within the crystal filter bandwidth, thus limiting the maximum scan frequency. Use of a high frequency scan would require wider bandwidth or separate processing channel for the directional information.

4. The inverse receiver for a monopulse angle tracking system

A conventional monopulse angle tracking scheme normally requires three complete

10.3 Semi-active Radar Homing Heads

channels, which are sum channel, elevation difference channel, and azimuth channel. The gain and phase of the three channels must be consistent. The well known advantage of monopulse over conical scan, however, is significant, since it eliminates externally-generated amplitude fluctuations (such as propeller modulation, fading noise, or amplitude modulated jammers). The monopulse system extracts all the angular information simultaneously rather than requiring a period of time to determine the position of a source of signals, as a conical scan system does.

The inverse receiver permits the inherent capability of the monopulse and the basic simplicity of the conical scan systems to be combined into one. A block diagram of an inverse monopulse receiver is indicated in Figure 10.3-6. The spectrums are indicated in Figure 10.3-7.

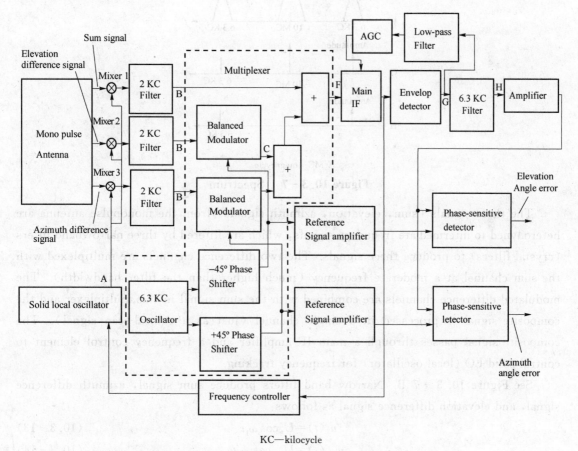

KC—kilocycle

Figure 10.3-6 Inverse monopulse receiver system

CHAPTER 10 Homing Guidance Systems

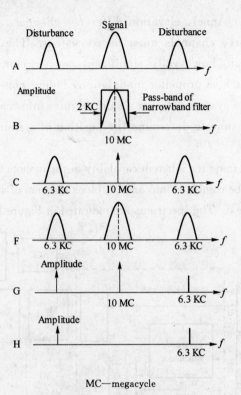

Figure 10.3-7 Spectrum

The three signals (sum, elevation, azimuth signals) from the monopulse antenna are heterodyned to intermediate frequency signals, which are filtered by three narrowband filters (crystal filters) to produce three signals. The two difference channels are multiplexed with the sum channel at a moderate frequency (much higher than the filter bandwidth). The modulated difference channels are combined with the sum signal in this multiplexer and the composite signal is processed in a single channel (just as a conical-scan signal). The composite signal passes through a main IF amplifier and a frequency control element to control solid LO (local oscillator) for frequency tracking.

See Figure 10.3-7 B. Narrow band filters produce sum signal, azimuth difference signal, and elevation difference signal as follows:

$$u_s(t) = U_s \cos \omega_I t \qquad (10.3-13)$$

$$u_{\Delta_A}(t) = U_{\Delta_A} \cos \omega_I t \qquad (10.3-14)$$

$$u_{\Delta_E}(t) = U_{\Delta_E} \cos \omega_I t \qquad (10.3-15)$$

where

U_s = amplitude of sum signal, represent echo energy,

U_{Δ_A} = proportional to azimuth angle error,

U_{Δ_E} = proportional to elevation angle error,

ω_I = intermediate frequency circular frequency.

By two opposite 45° phase shifters, the 6.3 KC oscillator signal (see Figure 10.3-6) is divided into two orthogonal reference signals as follows:

$$r_1(t) = A\sin \omega_M t \qquad (10.3-16)$$
$$r_2(t) = A\cos \omega_M t \qquad (10.3-17)$$

where $\omega_M = 2\pi f_M$, $f_M = 6.3$ KC.

The elevation angle error signal and the azimuth angle error signal are modulated by balanced modulator to upper sideband signals and lower sideband signals. The carrier signal is suppressed. See Figure 10.3-7 C. Two balanced modulators provide two signals as follows:

$$u_{M\Delta_E}(t) = K_M U_{\Delta_E} \cos \omega_I t \sin \omega_M t \qquad (10.3-18)$$
$$u_{M\Delta_A}(t) = K_M U_{\Delta_A} \cos \omega_I t \cos \omega_M t \qquad (10.3-19)$$

Sum of equations (10.3-13), (10.3-18) and (10.3-19) makes the composite signal as follows:

$$u_\Sigma(t) = U_s \left[1 + K_M \frac{U_\Delta}{U_s} \cos (\omega_M t - \psi) \right] \cos \omega_I t \qquad (10.3-20)$$

where

K_M = modulated coefficient,

$U_\Delta = \sqrt{U_{\Delta_E}^2 + U_{\Delta_A}^2}$,

$\psi = \arctan(U_{\Delta_E}/U_{\Delta_A})$.

The multiplexer output spectrum is shown in Figure 10.3-7 F. The composite signal is detected by an envelop detector to a DC signal and a sideband signal. The DC signal is sent to AGC. The sideband signal is sent to a 6.3 KC filter, which permits 6.3 KC signal to pass. See Figure 10.3-7 H. The 6.3 KC signal is amplified by a error amplifier to yield

$$u_d(t) = U_m \cos (\omega_M t - \psi) \qquad (10.3-21)$$

It is compared with two orthogonal references to produce two angle error signals, U_{EE}, U_{AE}, i.e.

$$U_{EE} = U_m \sin \psi \qquad (10.3-22)$$
$$U_{AE} = U_m \cos \psi \qquad (10.3-23)$$

The two error signals are used to control antenna servos and missile flight.

5. Radome (antenna cover)

In order to protect the homing head from the effect of outside circumstances, a radome (antenna cover) is mounted in front of the homing head. In the process of designing a radome, aerodynamic, structural, and electrical requirements should be considered. The radome should have the following advantages: small aerodynamic drag, high temperature tolerances, small loss when transmitting electromagnetic wave, and small aberration (or refraction) angle error, which arises from nonlinear distortions in the received energy as it passes through the antenna cover. The materials, such as glass, plastic, and fiberglass are generally used to make radomes.

When a homing head measures a target through a radome, due to the radome refraction to electromagnetic wave, the transmitting direction changes, i. e., the straight line transmission becomes the broken-line transmission. The result of refraction produces LOS error as shown in Figure 10.3 - 8.

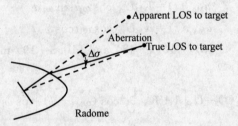

Figure 10.3 - 8 Radome refraction

For a given radome, the aberration is a nonlinear function of the look angle, $q - \theta_M = \varphi$ as shown in Figure 10.3 - 9.

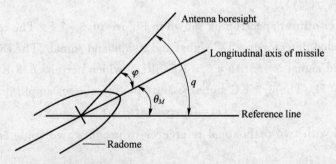

Figure 10.3 - 9 Geometrical relation

For different positions of the antenna, the aberrations are different, whose value may be positive, or negative. See Figure 10.3 - 10.

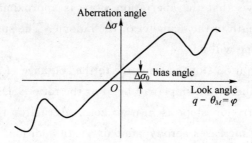

Figure 10.3 – 10 Aberration angle versus look angle

The aberration angle $\Delta\sigma$ is given by
$$\Delta\sigma = f(\varphi) \tag{10.3-24}$$
From this, the radome error slope is defined as
$$A = \frac{d\Delta\sigma}{d\varphi}$$
In general, $|A| \leqslant 0.05$.

Expanding the equation (10.3 – 24) to first-order Taylor series, the approximate expression of aberration angle $\Delta\sigma$ is
$$\Delta\sigma = \Delta\sigma_0 + A\varphi = \Delta\sigma_0 + A(q - \theta_M) \tag{10.3-25}$$
In view of the aberration effect, the angle q' as seen by the homing eye is given by
$$q' = q + \Delta\sigma = q + \Delta\sigma_0 + A(q - \theta_M) = \Delta\sigma_0 + (1+A)q - A\theta_M \tag{10.3-26}$$
From the equation (10.3 – 26) it can be seen that attitude angle θ_M produces a coupling to the guidance due to the radome error slope A. Assuming the transfer functions of the homing head and autopilot are formularized by second-order elements, the block diagram is plotted as shown in Figure 10.3 – 11. The aberration slope A provides a new feedback path. Through analysis of the block diagram, it is found that, since there is aberration slope, excess aberration slope A likely results in instability. Both positive and negative aberration slopes are harmful, but the negative aberration slope is much more dangerous.

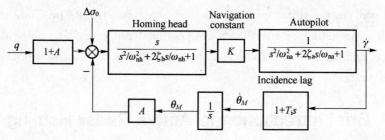

Figure 10.3 – 11 Parasitic feedback path arising from radome aberration

CHAPTER 10 Homing Guidance Systems

Research result shows that the aberration slope is approximately proportional to the ratio of the radio wave length to the diameter of the radome. The smaller the diameter is, the greater the aberration slope will be.

In addition, the aberration slope is sensitive to the acutance (L/D). The more acute the radome, the greater the aberration slope will be. For the ratio $L/D=1$, namely, the radome is a hemisphere, the aberration slope is near to zero. Actually, this case can not be used because the shape greatly increases aerodynamic drag. In addition, decreasing the thickness of the radome may reduce the aberration slope. We may also actively adopt compensating method to decrease the aberration effect. In advance, people measure aberrations for different position to form a compensating function, which cancels at real time the aberration actual effect by a control computer.

10.3.3 Target Illumination

Traditionally, the illuminator must continuously illuminate the target throughout the engagement. As it continuously tracks the target, it provides the illumination and the rear reference for missile guidance. Target illumination for a CW semi-active missile system can be provided by a CW tracking radar, a CW transmitter slaved to another tracking radar, or a pulse or pulse doppler tracking radar at another frequency with the CW illumination injected into the antenna system from a separate CW transmitter.

The most capable of these configurations is the CW tracking illuminator, as exemplified by the HARK surface-to-air system's High Power Illuminator (HPI). It is generally a two-dish radar because sufficient receiver-transmitter isolation cannot usually be achieved in a single-dish system. Alternatively, the illuminator can be the transmit-only portion of a radar slaved to a tracking radar—a mechanically scanned track-while-search (TWS) radar or a phased array which simultaneously maintains multiple target tracks with its electronically steered agile beam. In the third approach, where space constraints preclude use of separate antennas, such as in a fighter aircraft, a conventional pulse or PD (pulse doppler) radar tracks the target and the CW illumination is injected into the transmission port of the antenna from a separate transmitter.

10.4 Brief Introduction to Active Radar Homing Head

An active seeker is functionally the same as a semi-active seeker, with the exception that it carries its own illuminator. Besides adding the transmitter, the other main difference in the active seeker configuration is the elimination of the rear receiver, with the reference generated by offsetting the transmitter excitation (or drive) signal. It has an advantage of launched-and-forgotten (i. e. launch-and-leave). The launch-and-leave capability increases the protection for launch aircraft.

Figure 10.4 – 1 shows a pulse doppler active radar block diagram. Angle track still uses monopulse track (omitted in Figure 10.4 – 1). The figure only shows coherent receive-transmit system and inverse speed tracking loop. The active radar homing head employs an antenna for the transmitter and receiver. The timing controls the switch to isolate the transmitting and receiving signal. The system coherent reference is a local oscillator, which is a fixed frequency steady microwave source. By locked-phase, transmitting frequency is controlled by the local-oscillator, and follows the change of VCO (Voltage controlled oscillator) output. The seeker performs doppler tracking by means of controlling transmitted signal frequency. The speed tracking loop is closed by a target.

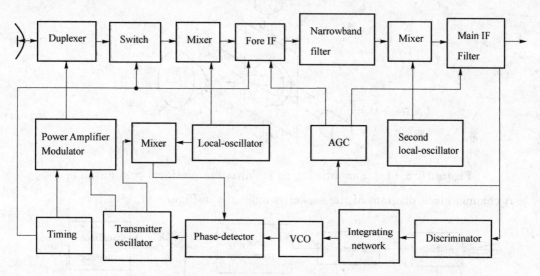

Figure 10.4 – 1　Pulse Doppler active radar block diagram

10.5　Antenna Boresight Stabilization and Track

Since the homing head is mounted on moving missile, in order to retain the boresight stabilized in space, the stabilization of some sort of gyroscope form in the head is required. If there is some base motion which tends to disturb the antenna, then pointing error will occur and be sensed by the gyroscope, and the servos will drive the antenna so as to bring the antenna back to its original position. For antenna boresight stabilization and track, two gyro stabilization schemes are widely used. One is a gyro stabilization platform scheme. The other is rate gyro feedback scheme.

10.5.1 Gyro Stabilization Platform Scheme

The antenna of the seeker is installed on a gyro stabilization platform. A simplified configuration is indicated in Figure 10.5-1. Using the stability of the gyroscope stabilizes the antenna boresight in inertial space to insulate disturbance from the airframe of a missile. Using the precession of the gyroscope controls the antenna boresight to track the target.

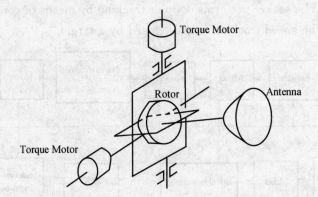

Figure 10.5-1 Simplified gyro stabilization platform configuration

A channel block diagram of the seeker is indicated in Figure 10.5-2.

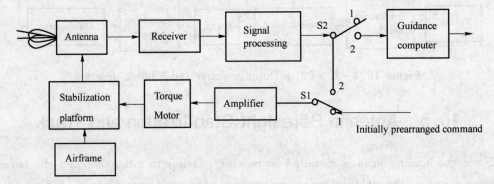

Figure 10.5-2 Gyro stabilization platform scheme

Before homing guidance, the antenna boresight needs to aim at a selected target. Switches S1 and S2 are shifted to positions 1. Through an amplifier, initially prearranged command is fed to the torque motor. The moment, which the torque motor produces, obliges the gyro to precess until the antenna boresight aims at the target. Then the switches S1 and S2 are shifted to positions 2. The receiver produces the angle error signals, which are used to control the antenna boresight for tracking target. At the same time, the error signal is used to guide the missile.

10.5.2 Rate Gyro Feedback Scheme

The antenna is fixed to a gimbaled platform, where two orthogonal rate gyros are mounted in a seeker as shown in Figure 10.5 - 3. The gimbaled platform is controlled by two servos.

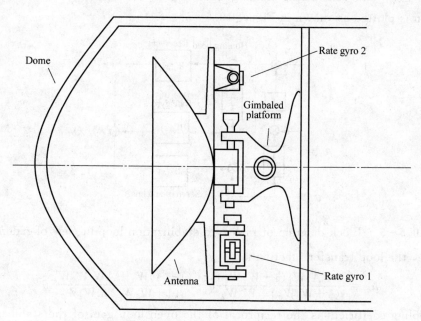

Figure 10.5 - 3 Rate gyro stabilization scheme

The stabilization loops are composed of the antenna servos and two rate gyros. In general, for the servos, there are two driving modes: hydraulic driving, and motor driving. Figure 10.5 - 4 indicates a vertical stabilization loop.

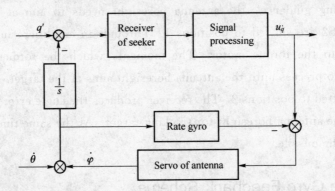

Figure 10.5-4 Block diagram of rate gyro stabilization loop

In view of radome's effect, the equation (10.3-26) is rewritten as

$$q' = \Delta\sigma_0 + (1+A)q - A\theta_M \tag{10.5-1}$$

If ignoring $\Delta\sigma_0$, and considering homing head noise, a block diagram of the rate gyro stabilization is plotted as shown in Figure 10.5-5.

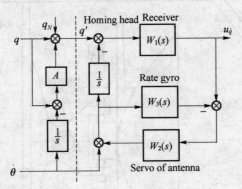

Figure 10.5-5 Block diagram of rate gyro stabilization loop in view of radome effect

Hence, the loop transfer function is

$$u_{\dot{q}} = \frac{(1+A)W_1(1+W_2W_3)}{s(1+W_2W_3)+W_1W_2}\dot{q} - \frac{(1+A)W_1+AW_1W_2W_3}{s(1+W_2W_3)+W_1W_2}\dot{\theta} \tag{10.5-2}$$

The uncoupling coefficient is the reciprocal of the open loop gain of the stabilization loop, namely,

$$k = \frac{1}{K_{W_2}K_{W_3}} \tag{10.5-3}$$

where K_{W_2} is gain of $W_2(s)$, and K_{W_3} is gain of $W_3(s)$. Being desirous to improve uncoupling ability, we increase open loop gain of the stabilization loop.

10.5.3 Noises Acting on Homing Heads

Noises acting on homing head result in stochastic error of angle measurement. The external noises, which enter a homing head, come from jamming, or natural refection of target. Let us put aside electronic countermeasures, and briefly discuss natural disturbances. A target, such as an aircraft, has complex structures and the configuration, every element of its surface is scattering source of radio frequency signals, and moreover, there are pitching, yawing, and rolling motions at any moment. These make reflecting surfaces produce stochastic motions relative to radar, as a result, arousing stochastic changes of the echoes, i.e. target noises. The target noises include amplitude fluctuation noise and angular noise.

With target motions, the amplitude fluctuation of the echo occurs. The amplitude fluctuations have greater effect on a conical scan seeker. It follows Rayleigh distribution. Figure 10.5-6 illustrates an example of spectral density of amplitude fluctuation. Spectral density peaks generally concentrate on the frequency range from 4 Hz to 10 Hz. Referring to Figure 10.5-6, if increasing scanning frequency, away from the peak frequency, this may greatly reduce the effect of the amplitude fluctuation. In theory, the amplitude fluctuation has no effect on a monopulse seeker, because the amplitude fluctuations in two difference channels are completely cancelled by that of the sum channel. In practice, however, since three channels can not complete consistent, the amplitude fluctuation effect still exists.

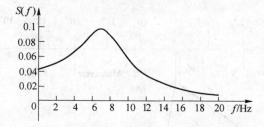

Figure 10.5-6 Spectral density of amplitude fluctuation of target echoes

Angular noise is sometimes known as angular glint noise, which is the fluctuation of the wave front of reflected wave. It is main error source of angle measurement. Its effect aggravates as the target approaches. The mean square deviation of angular fluctuation is estimated by

$$\sigma = (0.15 \sim 0.33) \frac{L}{R} \qquad (10.5-4)$$

where

L = target figure length, such as the wing span of an aircraft.

R = missile-target distance.

Inner noises of homing head include two types: ① receiver noise, and ② servo noise. The receiver noise belongs to thermal noise. With the reduction of the missile-target distance, and the signal-noise ratio increases, its effect decreases. The servo noise of a seeker has relation with design and fixed techniques, but no relation with missile-target distance and target reflection.

10.6 Infrared Seekers

10.6.1 Infrared Radiation

Any object with temperature greater than absolute zero K radiates infrared. Specially, moving military equipments, such as tanks, vehicles, ships, aircraft and so forth, are often strong infrared radiation sources because they have high temperature parts. Molecules, which compose object, run to generate thermal radiation. Physical essence of infrared is thermal radiation. Infrared is one electromagnetic wave, whose wavelength is 0.75~1 000 μm. Infrared spectrum is between visible light and millimeter as shown in Figure 10.6-1.

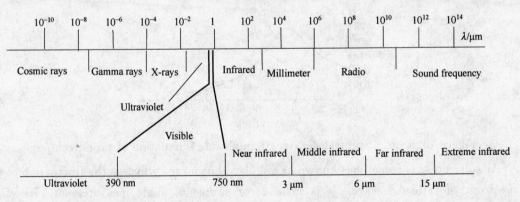

Figure 10.6-1 Electromagnetic spectrum

In general, infrared is divided into four bands: 0.75~3 μm near infrared (NIR), 3~6 μm middle infrared (MIR), 6~15 μm far infrared (FIR), and 15~1 000 μm extreme infrared (XIR).

Like visible light, infrared also transmits at light velocity. It satisfies the following relation:
$$\lambda f = c$$
where
λ = wavelength,
f = frequency,
c = light velocity, $2.997\,924\,58 \pm 0.000\,000\,012 \times 10^8$ m/s $\approx 3.00 \times 10^8$ m/s.

It also has phenomena of reflection, refraction, interference, diffraction, and polarization. For eyes, infrared is invisible, so infrared is received by infrared detectors. Photon energy of infrared is less than visible light.

When infrared passes through atmosphere, the radiant flux is attenuated since some radiant energy is absorbed by atmosphere. The transmittance of infrared radiation through atmosphere is expressed as
$$\tau = e^{-\sigma x}$$
where
σ = extinction coefficient,
x = path length.

When the infrared passes through atmosphere, there are several regions of high transmission, called atmospheric windows. Regions of high absorption intervene in them. In common use, the atmospheric windows are $0.75 \sim 1.1$ μm, $3 \sim 5$ μm and $8 \sim 14$ μm.

10.6.2 Introduction to Detectors

The detectors of infrared are classified into thermal detectors and photon detectors. The thermal detector works based on thermal effect of infrared radiation. The photon detector works based on the fact that when incident photon stream radiates onto the infrared detector, the resistance variation, or optical voltage is generated. Therefore the response of the thermal detector is proportional to energy of absorption, but that of the photon detector is proportional to photon number of absorption. For infrared seekers, the photon detectors are commonly used, since its sensibility and response velocity are superior to those of the thermal detector. The photon detectors include three types as follows.

1. Photoconductive detector

When infrared photon streams radiate onto the device, the carriers are excited so that the resistance decreases, or conductance increases. The detector using this phenomenon is called photoconductive detector. The sort of detector is also known as photoresistance. Lead sulfide (PbS), indium antimonide (InSb), germanium doping of mercury, germanium doping of gold, and germanium doping of copper belong to this type.

2. Photovoltaic detector

When infrared photon streams radiate onto the device, the voltage is generated. Using this phenomenon can detect infrared. In general, there are indium arsenide (InAs), indium antimonide (InSb), and mercury-cadmium-telluride (HgCdTe) detectors.

3. Photoelectromagnetic detector

It consists of a slice of intrinsic semiconductor material and a magnet. When infrared photon streams radiate onto the device, the electron-hole pairs are produced, and separated by the magnetic field to generate electromotive force. The detectors do not need cooling. Since the detectivity of the device is lower than those of the photoconductive detector and the photovoltaic detector, it is rarely applied.

At present, lead sulfide (PbS), indium antimonide (InSb), and mercury-cadmium-telluride (HgCdTe) detectors are widely used.

(1) Performance parameters of infrared

1) Noise equivalent power (NEP)

NEP is defined as radiant power of incident detector when its output power is equal to noise power. It is given by

$$\text{NEP} = \frac{HA_d}{V_s/V_n} \qquad (10.6-1)$$

where

$H=$ irradiance, radiant flux emitted unit area,

$A_d=$ photosensitive area of detector,

$V_s=$ root-mean-square of detector output signal,

$V_n=$ root-mean-square of noise voltage of detector output signal.

NEP represents an ability which the detector can detect minimum radiation power. The smaller the value, the stronger the ability will be.

2) Detectivity (D), and normalized detectivity (D^*)

D is defined as the reciprocal of the NEP, i. e.

$$D = 1/\text{NEP} \qquad (10.6-2)$$

The greater the value of D, the stronger detectability will be. In view of effects of area of detector, A_d and bandwidth Δf, the detectivity is normalized as

$$D^* = D(A_d \Delta f)^{1/2} = \frac{(A_d \Delta f)^{1/2}}{\text{NEP}} \qquad (10.6-3)$$

3) Noise equivalent temperature difference(NETD)

When an infrared system observes the tested figure, if the ratio of the peak signal of detector output to root-mean-square of noise voltage of detector output signal is 1, the temperature difference between the target and the background is called noise equivalent temperature difference.

4) Minimum recognition temperature difference(MRTD)

A figure with four strips is used to test MRTD. The strip length is 7 times the strip width. The blackbodies intervene in the strips. The initial temperature difference is zero, and then the temperature difference is gradually increased until the strips are completely recognized from a display. At this time, the difference is known as MRTD.

(2) Refrigerator of detector

Infrared detectors are mostly narrowband semiconductor. In normal temperature, thermal noise is great, so detectors work in cryogenic state. The range is from 125 K to absolute zero K. Hence, they need special refrigerator to normally work. In infrared system, there are commonly Dewar, gas throttling refrigerator, Stirling refrigerator, Joule-Thomson refrigerator and semiconductor refrigerator. Since the volume of onboard seeker is restricted, the micromation of refrigerator is required. Manufacture technologies of micro-refrigerators are complicated. At present, the micro-refrigerator is a key factor to infrared systems.

10.6.3 Infrared Point Source Seekers

The infrared point source seeker tracks a target based on an infrared image point of the target. It belongs to passive seekers. It has an advantage of launched-and-forgotten (i. e. launch-and-leave). It is widely used in surface-air, air-air, anti-ship, anti-tank missiles. The elements of the seeker are shown in Figure 10.6-2.

The infrared seeker includes an optical system, a scan modulation, a refrigerator, a signal processor, a gyro servo, and a command generator. The optical system receives infrared radiation signal of a target. After modulated by the modulator, the optical signal

CHAPTER 10 Homing Guidance Systems

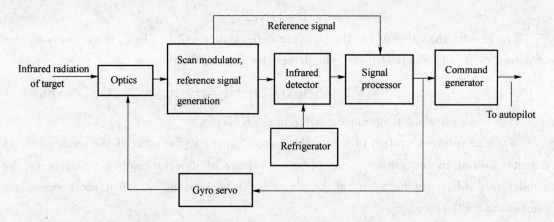

Figure 10.6 – 2 Block diagram of elements of infrared seeker

with orientation information is produced. The optical signal is converted into electronic signal by the infrared detector. The electronic signal is filtered, amplified, and demodulated by the signal processor to obtain angle error signal. The angle error is sent to the gyro servo to drive the optical system precession. As a result, the boresight tracks the target. At the same time, a guidance signal is generated.

According to different modulation mode, the infrared point source seeker is classified into the seeker with reticle and the seeker without reticle.

1. Seeker with reticle

In the seeker, a specialized rotating reticle performs space filter. Using the reticle can extracts target information from backgrounds. Figure 10.6 – 3 shows a basic reticle. The figure of the reticle consists of fan-shaped regions, where bright portions and dark portions occur alternately. The bright portions permit infrared to pass through, while the dark portions prohibit infrared to penetrate. For a detail discussion about reticle, please refer to the Reference [5].

Figure 10.6 – 3 Basic reticle

2. Seeker without reticle

At present, infrared point source seeker without reticle is widely used. Infrared optical receiving device includes an irdome, a primary mirror, a planar mirror, a lens, and an optical filter as shown in Figure 10.6 – 4.

290

10.6 Infrared Seekers

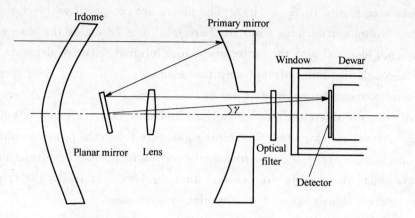

Figure 10.6-4 Infrared optical receiving device

In optical path, since the planar mirror is obliquely placed, the reflected ray has an angle γ compared with light axis. The planar mirror rotates with a gyro rotator. Focalized target image point rotates. The sensitive surface of InSb detector is located at the focal plane of the optical system. Four detector elements (as seen in Figure 10.6-5) are distributed orthogonally.

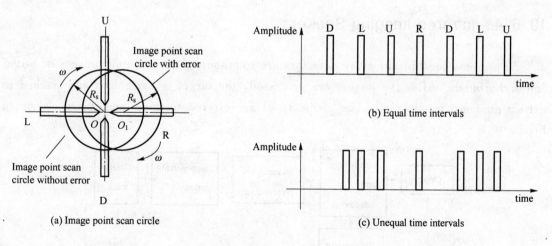

Figure 10.6-5 Optical modulation system without reticle

The image point rotates at angular speed ω and radius R_s. When a target locates at the boresight, the center of the scan circle of the image point is identical with the center of the four infrared sensitive elements, and the image point passes through the four infrared sensitive elements at same time intervals. The detector generates a series of the same time interval pulses (see Figure 10.6 – 5 (b)). The pulses are compared with references to zero signals. When the target deviates from the boresight, the center of the scan circle of the image point is not identical with the center of the four infrared sensitive elements. The image point scans through the four infrared sensitive elements at unequal time intervals. The detector generates unequal time interval pulses (see Figure 10.6 – 5 (c)). After these pulses are compared with the reference pulses, and manipulated, two angle errors (elevation and azimuth angle errors) are extracted. The error signals are fed to the power amplifiers, which supply the currents to excite the electromagnetic valves of pneumatic torque motors. The torque motors oblige the gyro to precess such that the boresight tracks the target. At the same time, the error signals are sent to autopilot for guidance.

Magnetoelectric manner is used to generate the references. A small alnico is fixed on the rotator over the planar mirror, while four inducing coils are mounted orthogonally on the gyro stator. Distributed direction of four inducing coils is consistent with those of four infrared sensitive elements. When the alnico rotates, four coils produce inductive electromotive forces. The phase difference of neighboring signals is 90°. After rectified and chopped, the signals are used as reference signals.

10.6.4 Infrared Imaging Seekers

At present, focal-planar array detectors are commonly used to obtain images of target and backgrounds. After the images are processed, the target is recognized and tracked to extract guidance intelligence. The elements of an infrared imaging seeker are shown in Figure 10.6 – 6.

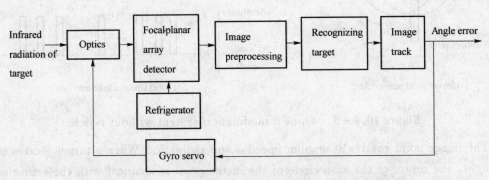

Figure 10.6 – 6 Elements of infrared imaging seeker

Infrared imaging seekers have advantages as follows:
① Have stronger anti-jamming ability,
② Effectively identify visible disguise,
③ Operate day and night,
④ Realize launched-and-forgotten.

The disadvantage is that the infrared seekers are susceptible to atmospheric conditions, such as moist atmosphere, fog, cloud, and so forth.

1. Focal-planar array detector

In infrared imaging seekers, middle infrared, $3 \sim 5$ μm or far infrared, $8 \sim 12$ μm is commonly used. In the past, as restricted by technology level, only single element detector or line array detector were made. If the line array detector is used to make imaging seeker, in order to obtain planar image, one dimensional scan mechanics must be added. As a result, the seeker becomes more complicated, and its gravity, volume, and power consumption increase. With the development of detector techniques, focal-planar array detectors are extensively applied. In general, the detectors have different scales such as 128×128, 320×240, 640×480 elements. As more elements of focal-planar array occur, it is easier to obtain clearer images, but the cost increases.

2. Image tracker

The image tracker can measure angular errors from the light axis to be fed to servos. The servos drive the light axis to track a target. Through a monitor, one can watch tracking states. In general, image tracking methods include centroid tracking, edge tracking, and correlation tracking.

The angular errors are given by

$$\Delta A = \frac{\Delta x}{f \cos E} \quad (10.6-4)$$

$$\Delta E = \frac{\Delta y}{f} \quad (10.6-5)$$

where

ΔA = angular error of azimuth,
Δx = deviation from the light axis in horizontal direction of imaging plane,
f = foci,
E = elevation angle,

ΔE = angular error of elevation,

Δy = deviation from the light axis in vertical direction of imaging plane.

10.7 Homing Guidance System Design

10.7.1 Homing Guidance Geometrical Relation

Figure 10.7-1 illustrates geometrical relation between a missile and a target in longitudinal plane.

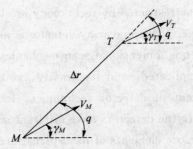

Figure 10.7-1 Missile-target geometrical relation

From Figure 10.7-1, the angular velocity perpendicular to LOS is

$$\omega_{LOS} = \dot{q} = \frac{V_M \sin(q - \gamma_M) - V_T \sin(q - \gamma_T)}{\Delta r} \quad (10.7-1)$$

If $(q - \gamma_M)$ and $(q - \gamma_T)$ are assumed to be small, then equation (10.7-1) is linearized. Thus

$$\dot{q} = \frac{V_M(q - \gamma_M) - V_T(q - \gamma_T)}{\Delta r} \quad (10.7-2)$$

Taking Laplace transformation of equation (10.7-2) yields

$$q = \frac{\frac{V_T}{\Delta r}(\gamma_T - q)}{s - \frac{V_M}{\Delta r}} - \frac{\frac{V_M}{\Delta r}\gamma_M}{s - \frac{V_M}{\Delta r}} \quad (10.7-3)$$

q acts as the input of the seeker. The output of the seeker is the measured LOS rate ω, which is the input to the guidance computer. The guidance computer generates the acceleration command for the autopilot of the missile. The missile's acceleration is divided by the missile's velocity to generate $\dot{\gamma}_M$, which is integrated to generate γ_M. Figure 10.7-2 shows the block diagram of the guidance system. From Figure 10.7-2, it can be seen that the feedback path introduces a pole in the right half s-plane. This should be sufficiently considered in design.

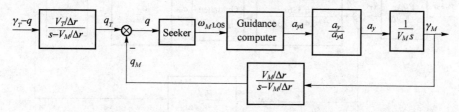

Figure 10.7-2 Block diagram of homing guidance system

10.7.2 Example of Homing Guidance System Design

As mentioned earlier for proportional navigation guidance, the turning rate of a missile is proportional to the LOS angular velocity. The ratio of the missile turning rate to the LOS angular velocity is called the proportional navigation constant N. The N usually ranges from 2 to 6. The seeker measures the LOS angular velocity. A simplified seeker block diagram is shown in Figure 10.7-3.

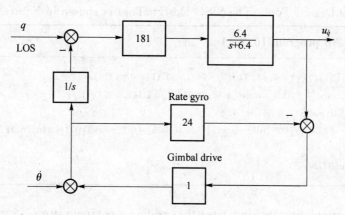

Figure 10.7-3 Simplified seeker block diagram

In primary design, ignoring the missile coupling effect, then

$$\frac{u_{\dot{q}}}{q} = \frac{181 \times \dfrac{6.4}{s+6.4}}{1+24+181 \times \dfrac{6.4}{s+6.4} \times \dfrac{1}{s}} = \frac{46.336 s}{s^2 + 6.4 s + 46.336} \tag{10.7-4}$$

The block diagram of proportional navigation guidance can be drawn as shown in Figure 10.7-4.

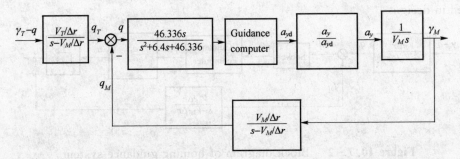

Figure 10.7-4 Block diagram of proportional navigation guidance

The a_y/a_{yd} transfer function was given in equation (8.5-6), which is repeated here for convenience:

$$\frac{a_y}{a_{yd}} = \frac{113\,400(-0.36s^2 - 1.08s + 1\,320)}{(s+21.4)(s^2+18.94s+254.8)(s^2+145.6s+27\,450)} \quad (10.7-5)$$

For this case, $V_M = 500$ m/s, $\Delta r = 5\,000$ m.

If the navigation constant $N = 5$, then the gain is $5 \times 500 = 2\,500$. Use RLTOOL of MATLAB (SISO Design Tool) to design guidance loop. MATLAB program 10.7-1 generates the SISO Design Tool. The SISO Design Tool is shown in Figure 10.7-5.

```
MATLAB program 10.7-1

>> n=[113400*(-0.36) 113400*(-1.08) 113400*1320];
>> d=conv([0 1 21.4],conv([1 18.94 254.8],[1 145.6 27450]));
>> G=tf(n,d);
>> n1=[43.336/5000];d1=conv([1 6.4 46.336],[0 1 -0.1]);G1=tf(n1,d1);
>> G2=G*G1;
>> rltool(G2);
```

The result of simulation test indicates that adding a compensator $G_{comp} = \dfrac{2\,500(1+3.5s)}{(1+0.73s)}$ will produce satisfactory frequency domain indexes. Figure 10.7-5 shows the gain margin and phase margin as follows:

$$Gm = 6.03 \text{ dB (at } 5.26 \text{ rad/s)}, \quad Pm = 83.8° \text{ (at } 1.93 \text{ rad/s)}$$

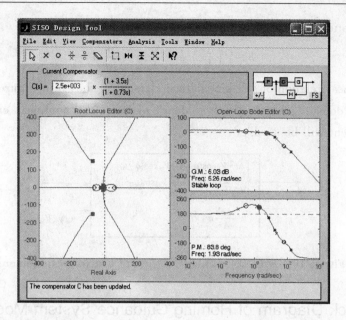

Figure 10.7-5 SISO Design Tool of the homing guidance loop

10.8 Homing Guidance System Model

10.8.1 Nonlinear Kinematical Element of Homing Guidance

The kinematical element expresses missile-target relative motion relation, which forms a closed loop of homing guidance by feedback missile-target relative motion information to a seeker. Here, the equation (10.7-1) is rewritten as follows

$$\dot{q} = \frac{V_M \sin(q-\gamma_M) - V_T \sin(q-\gamma_T)}{\Delta r} \qquad (10.8-1)$$

In addition, from Figure 10.7-1,

$$\Delta \dot{r} = V_T \cos(q-\gamma_T) - V_M \sin(q-\gamma_M) \qquad (10.8-2)$$

For convenience, the miss distance formula (7.3-6) is rewritten as

$$h = V t_{go}^2 \omega_{LOS} \qquad (10.8-3)$$

We might as well consider the time-to-go is approximately

$$t_{go} = \frac{\Delta r}{V} \qquad (10.8-4)$$

297

Substituting equation (10.8-4) into equation (10.8-3) yields

$$h = V\left(\frac{\Delta r}{V}\right)^2 \dot{q} = \frac{\Delta r^2}{V} \dot{q} = \frac{\Delta r^2}{\Delta \dot{r}} \dot{q} \qquad (10.8-5)$$

Thus, the nonlinear relation of the kinematical element is shown in Figure 10.8-1. When performing time domain analysis for a guidance system, the nonlinear expressions of the kinematical element will be used.

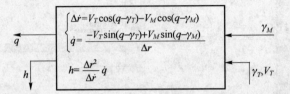

Figure 10.8-1   Nonlinear relation of kinematical element

10.8.2   Block Diagram of Homing Guidance System Model

Homing guidance system mainly includes a seeker, a generator of guidance command (include a filter and a compensator), an autopilot, and a missile-target relative motion element as shown in Figure 10.8-2.

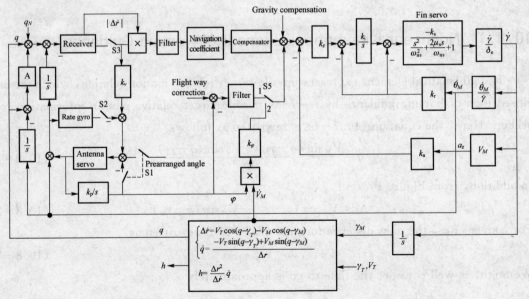

Figure 10.8-2   Block diagram of homing guidance system

Before the missile is launched, a lead angle of the seeker antenna direction is sometimes preset such that after the missile is launched, the target can accurately fall within the antenna beam. This ensures the seeker to rapidly acquire the target. Switch S1 is on. Switches S2 and S3 are off. The servo of the seeker is a typical angular position control system. The prearranged lead angel is provided by the fire control radar on ground or aircraft.

After the missile is launched, prearranged switch S1 and tracking switch S3 are off, and stabilization switch S2 is on. In general, within 2 seconds after launching, since bad ground clutters and forward wave leakage effect, the echo signal-to-noise ratio is very low such that the seeker can not acquire the target. At this time, the tracking loop is on open loop state. Only the stabilization loop works to hold the antenna direction stabilization in inertial space.

The seeker begins to sweep. Once the seeker acquires the target, tracking switch S3 is on. The seeker continuously detects and tracks the target, and yields guidance commands.

When an air-to-air missile is launched, the longitudinal axis of missile does not often aim at demanded direction. After the missile is launched, it is necessary to correct initial error of flight way. At this time switch S5 is on at position 1.

Since the missile speed change results in the LOS rate change, Switch S5 is on at position 2 to compensate the effect of tangent acceleration of a missile.

References

[1] Garnell P. Guided Weapon Control Systems. Oxford:Pergamon Press,1980.
[2] John H Blakelock. Automatic Control of Aircraft and Missiles. 2nd ed. New York:Wiley, 1991.
[3] George M Siouris. Missile Guidance and Control. New York:Spring-Verlag, Inc. ,2004.
[4] Skolnik M I. Radar handbook. 2nd ed. New York: McGram-Hill, 1990.
[5] Richard D, Hudson Jr. Infrared system engineering. New York:John Wiley & Sons, Inc. , 1969.
[6] Ivanov A. Simi-active Radar Guidance. Microwave Journal, 1983,9:105-120.
[7] 穆虹. 防空导弹雷达导引头设计. 北京:宇航出版社, 1996.

CHAPTER 11
Hardware-in-the-loop Simulation of Guidance and Control System

11.1 Functions of Hardware-in-the-loop Simulation

The hardware-in-the-loop simulation of guidance and control system means to consider guidance and control element functions or performances and mutually connecting relations through physical devices, physical environments and mathematical models. Making use of onboard devices and building battlefield environments as realistic as possible, people test the guidance and control system of missile.

In summary, the functions of hardware-in-the-loop simulation mainly include: ① verify guidance and control system functions in most close to possible war environment, ② consider some parts and some elements' effects on guidance and control system, ③ verify performances of subsystems and cooperation of devices, ④ evaluate mathematical models of the guidance and control system.

11.2 Hardware-in-the-loop Simulation System

11.2.1 Hardware-in-the-loop Simulation System Components

To a great extent, the hardware-in-the-loop simulation system components depend on the guidance system, which the missile employs, and the manner of detecting target. Hereby, there are normally the command guidance, radio frequency guidance, infrared imaging guidance, and optical-electronic guidance hardware-in-the-loop simulation systems.

Although exist some differences in them, they still have some common points. The hardware-in-the-loop simulation includes participating equipment, such as infrared seeker, autopilot and other missile-borne devices, simulation equipment, such as three axes flight table, load simulator, acceleration simulator, target simulator and so on. For example, Figure 11.2 – 1 indicates block diagram of the hardware-in-the-loop simulation system. The simulator of infrared target simulates infrared wave of targets, which is detected by an onboard infrared seeker. The seeker tracks the target, simultaneously generates two error signals to the onboard guidance computer. After selecting guidance law, the guidance computer generates control commands to the autopilots of the missile, which steer actuators to deflect control surfaces of missile. The angles, which the control surfaces deflect, are sampled, and then are fed to the simulation computer. The simulation computer computes the aerodynamic forces, the aerodynamic moments, the position and the attitude of the missile, and the target motion information. The attitude information is sent to three-axis flight simulator. The target motion information is sent to the two-axis target motion simulator and the imaging computer of the target. With the changes of the missile and target position and attitude, the guidance intelligence, which is updated continuously, corrects missile flight path.

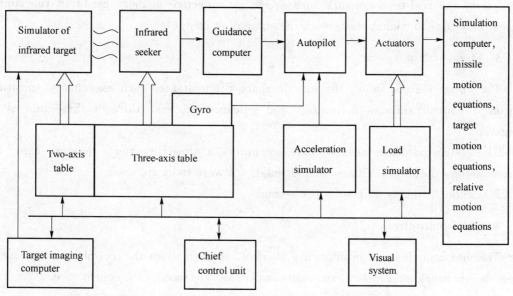

Figure 11.2 – 1 Block diagram of hardware-in-the-loop simulation system of infrared imaging homing guidance system

11.2.2 Subsystem Functions

1. Flight simulator

Three-axis flight simulator (three-axis table) is used to simulate missile attitude motion. The three-axis simulator consists of a table, control cabinet, and monitor. The displacement gyroscopes, rate gyroscopes, and a seeker are mounted on a flight simulator, or a seeker is mounted on another flight simulator. The control computer of the flight simulator receives the simulation command from the simulation computer. Because real-time is required for simulation, the real-time network card (such as reflective memory card) is needed to communicate with the simulation computer.

2. The moment load simulator

It simulates the aerodynamic load, which acts on control surfaces of missile. The moment load simulator of missile includes a four-channel loading table, control cabinet, monitor, and hydraulic pump source. The control computer of the loading table receives the hinge moment signal from the simulation computer. Because real-time is required for simulation, the real-time network card (such as reflective memory card) in the control cabinet is needed to communicate with the simulation computer.

3. Visual system

The visual system shows the missile space flight states, such as velocity, position, attitudes, control surfaces deflection, and guidance process, through three-dimensional animation.

The system hardware consists of a computer, a projector, and a network card. For programming, the Vega, VTree, and OpenGL software tools are used.

It is enough to play 30 frames per second.

4. Chief controller

The chief controller has an interactive interface, through which the control commands can be given. It may initialize the system, show data and curves, and monitor the system state.

5. Acceleration simulator

The acceleration simulator can construct environment of acceleration simulation, where the accelerometer of the missile is mounted. The sensed acceleration value in the environment is fed to the autopilot of missile to accomplish control in closed loop.

It includes a table and controller. It receives commands from the simulation computer.

6. Simulation computer

The simulation computer performs calculations of six-degree-freedom dynamic equations and kinematical equations and provides control signals to simulation equipments.

The hardware includes CPU card, A/D card, D/A card, DI/DO card, RS—422 card, real-time network card. In general, its configuration is higher.

7. Communication system

Since the simulation system requires higher real-time, RTnet (real-time net) is normally employed. It is a real-time system based on the real-time network cards, and for transmission the optical fiber is used.

In market, there are three typical products about the real-time network cards as follows:

① RTnet from VMIC,
② SCRAM NET from SYSTRAN,
③ Broadcast Memory Network from SBS.

Generally, there are two typical structures as follows: ① "△"type, and ② "Y"type as shown in Figure 11.2 - 2.

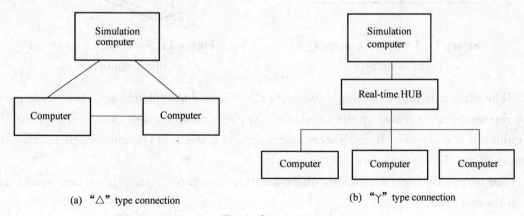

(a) "△" type connection (b) "Y" type connection

Figure 11.2 - 2   Typical communication structure

11.3 Simulation Equipments

11.3.1 Three-axis Flight Simulator

Figure 11.3-1 shows mechanical structure of three-axis flight simulator.

The three axes of flight simulator are inner gimbal axis x, middle gimbal axis y and outer gimbal axis z respectively. See Figure 11.3-2. According to the commands from the simulation computer, the motions about three axes are accurately controlled by a computer. As to driving modes, there are normally electrically driving mode, hydraulically driving mode, and electrically plus hydraulically driving mode. For hydraulic flight simulator, it needs a hydraulic source.

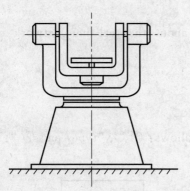

Figure 11.3-1 Three-axis flight simulator

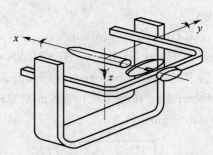

Figure 11.3-2 Three axes of flight simulator

The three-axis flight simulator simulates the missile's attitude change. If attitude gyros and rate gyros are mounted on the simulator, the gyros sense attitude change to be fed to the autopilot of the missile. If the seeker is mounted on the flight simulator, it follows the motion of the flight simulator.

For three-axis flight simulator, there are two structure forms: ① vertical style, and ② horizontal style.

1. Vertical style

Vertical style is as shown in Figure 11.3-3.

11.3 Simulation Equipments

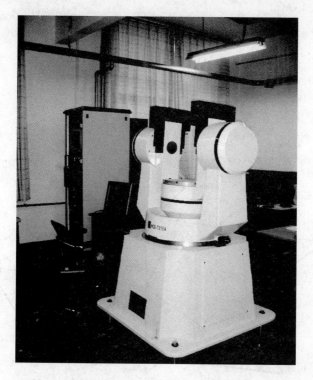

Figure 11.3-3 Vertical style

In vertical style three-axis flight simulator, the direction of the outer gimbal axis z is unchanged. The directions of the inner gimbal and middle gimbal axes change with the motion of the outer gimbal. Figure 11.3-4 shows the relation between the vertical style simulator motion and the missile body coordinate system.

The coordinate system $S-xyz$ is the missile body coordinate system. At the beginning, the three axes are mutually orthogonal, corresponding to Ox_E, Oy_E, and Oz_E. The outer, middle, and inner gimbals of the simulator rotate in turn for ψ, θ, and ϕ. As a result, the coordinate system $S-x_E y_E z_E$ is transformed onto the body coordinate system $S-xyz$. But finally, the three axes are located at Ox_{inner}, Oy_{mid}, and Oz_{outer}. The Oy_{mid} axis is perpendicular to the Ox_{inner} axis and the Oz_{outer} axis, but the Ox_{inner} axis is not perpendicular to the Oz_{outer} axis. Thus,

$$\begin{bmatrix} p \\ q \\ r \end{bmatrix} = \begin{bmatrix} \dot{\phi} \\ 0 \\ 0 \end{bmatrix} + \begin{bmatrix} 1 & 0 & 0 \\ 0 & \cos\phi & \sin\phi \\ 0 & -\sin\phi & \cos\phi \end{bmatrix} \begin{bmatrix} 0 \\ \dot{\theta} \\ 0 \end{bmatrix} +$$

CHAPTER 11 Hardware-in-the-loop Simulation of Guidance and Control System

$$\begin{bmatrix} 1 & 0 & 0 \\ 0 & \cos\phi & \sin\phi \\ 0 & -\sin\phi & \cos\phi \end{bmatrix} \begin{bmatrix} \cos\theta & 0 & -\sin\theta \\ 0 & 1 & 0 \\ \sin\theta & 0 & \cos\theta \end{bmatrix} \begin{bmatrix} 0 \\ 0 \\ \dot\psi \end{bmatrix} =$$

$$\begin{bmatrix} 1 & 0 & -\sin\theta \\ 0 & \cos\phi & \cos\theta\sin\phi \\ 0 & -\sin\phi & \cos\theta\cos\phi \end{bmatrix} \begin{bmatrix} \dot\phi \\ \dot\theta \\ \dot\psi \end{bmatrix} \quad (11.3-1)$$

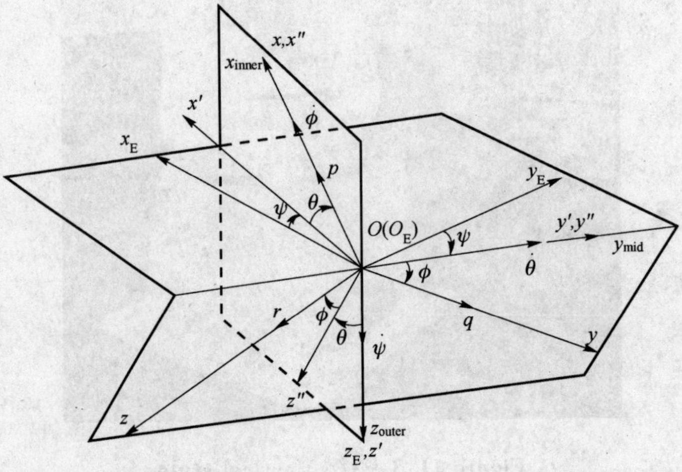

Figure 11.3-4 Relation between the vertical style simulator motion and the missile body coordinate system

The conversion matrix is given by

$$\begin{bmatrix} \dot\phi \\ \dot\theta \\ \dot\psi \end{bmatrix} = \begin{bmatrix} 1 & \sin\phi\tan\theta & \cos\phi\tan\theta \\ 0 & \cos\phi & -\sin\phi \\ 0 & \sin\phi\sec\theta & \cos\phi\sec\theta \end{bmatrix} \begin{bmatrix} p \\ q \\ r \end{bmatrix} \quad (11.3-2)$$

where

p, q, and r are rolling rate, pitching rate, and yawing rate of missile respectively,

ψ, θ, and ϕ are rotating angles of the simulator,

$\dot\phi$, $\dot\theta$, and $\dot\psi$ are used to control the inner gimbal, middle gimbal, and the outer gimbal axes respectively.

2. **Horizontal style**

Horizontal style is as shown in Figure 11.3-5.

11.3 Simulation Equipments

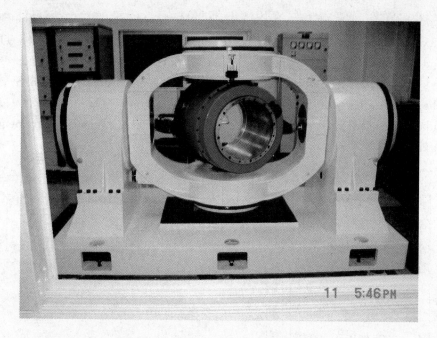

Figure 11.3-5   Horizontal style

In horizontal style three-axis flight simulator, the direction of the outer gimbal axis y is unchanged. The directions of the inner and middle gimbal axes change with the motion of the outer gimbal. Figure 11.3-6 shows the relation between the horizontal style simulator motion and the missile body coordinate system.

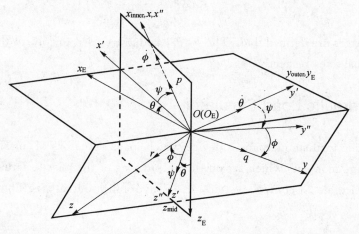

Figure 11.3-6   Relation between the vertical style simulator motion and the missile body coordinate system

The coordinate system $S—xyz$ is the missile body coordinate system. At the beginning, the three axes are mutually orthogonal, corresponding to Ox_E, Oy_E, and Oz_E. The outer, middle, and inner gimbals of the simulator rotate in turn for θ, ψ, and ϕ. As a result, The coordinate system $S—x_E y_E z_E$ is transformed onto the body coordinate system $S—xyz$. But finally, the three axes are located at Ox_{inner}, Oy_{outer}, and Oz_{mid}. The Oz_{mid} axis is perpendicular to the Ox_{inner} axis and the Oy_{outer} axis, but the Ox_{inner} axis is not perpendicular to the Oy_{mid} axis.

Thus,

$$\begin{bmatrix} p \\ q \\ r \end{bmatrix} = \begin{bmatrix} \dot{\phi} \\ 0 \\ 0 \end{bmatrix} + \begin{bmatrix} 1 & 0 & 0 \\ 0 & \cos\phi & \sin\phi \\ 0 & -\sin\phi & \cos\phi \end{bmatrix} \begin{bmatrix} \cos\psi & \sin\psi & 0 \\ -\sin\psi & \cos\psi & 0 \\ 0 & 0 & 1 \end{bmatrix} \begin{bmatrix} 0 \\ \dot{\theta} \\ 0 \end{bmatrix} + \begin{bmatrix} 1 & 0 & 0 \\ 0 & \cos\phi & \sin\phi \\ 0 & -\sin\phi & \cos\phi \end{bmatrix} \begin{bmatrix} 0 \\ 0 \\ \dot{\psi} \end{bmatrix} =$$

$$\begin{bmatrix} 1 & \sin\psi & 0 \\ 0 & \cos\psi\cos\phi & \sin\phi \\ 0 & -\cos\psi\sin\phi & \cos\phi \end{bmatrix} \begin{bmatrix} \dot{\phi} \\ \dot{\theta} \\ \dot{\psi} \end{bmatrix} \quad (11.3-3)$$

The conversion matrix is given by

$$\begin{bmatrix} \dot{\phi} \\ \dot{\theta} \\ \dot{\psi} \end{bmatrix} = \begin{bmatrix} 1 & -\cos\phi\tan\psi & \sin\phi\tan\psi \\ 0 & \cos\phi & -\sin\phi \\ 0 & \sin\phi & \cos\phi \end{bmatrix} \begin{bmatrix} p \\ q \\ r \end{bmatrix} \quad (11.3-4)$$

where, these letters are defined ibid. The $\dot{\phi}$, $\dot{\theta}$, and $\dot{\psi}$ are used to control the inner, outer, and middle gimbal axes respectively.

11.3.2 Hydraulic Load Simulator

The hydraulic load simulator is a device which loads actuators of missile. It is used to simulate the hinge moments which act on the output axis of the missile actuators. It consists of a four-channel table, a control system, and a hydraulic source. See Figure 11.3-7 and Figure 11.3-8.

11.3 Simulation Equipments

Figure 11.3 − 7   Hydraulic load simulator

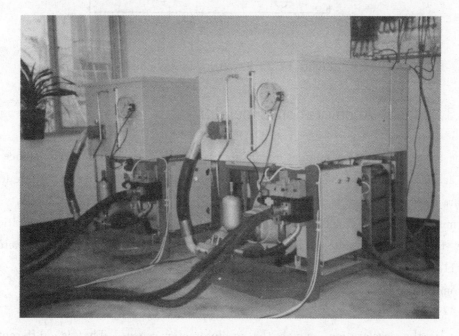

Figure 11.3 − 8   Hydraulic source

Figure 11.3-9 shows structure of a loading channel. A loading channel consists of a servo valve, a hydraulic motor, a torque sensor, an angle sensor, and inertial pieces. The hydraulic motor acts as a driving part. The torque sensor is a feedback part, which forms a closed loop torque control system. In order to eliminate (or reduce) the extra torque, the angle sensor is set up. Even though the given torque is zero, output torque is not zero due to the actuator motion. This output torque is called the extra torque. Based on investigation, it is found that the extra torque is proportional to the angular velocity. Thus, the signal, which is proportional to the angular speed, is used to eliminate the extra torque.

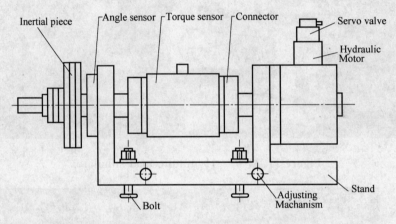

Figure 11.3-9   A loading channel

11.3.3 Linear Acceleration Simulator

The linear acceleration simulator can generates the acceleration circumstance for the missile accelerometer. It generally includes a constant speed rotation table, where follow-up tables are fixed, as shown in Figure 11.3-10.

A missile accelerometer is mounted on the follow-up table. Changing the rotation angle of the follow-up table may change the sensitive direction of accelerometer. See Figure 11.3-11.

From Figure 11.3-11,

$$a_n = R\omega^2 \sin \alpha$$

where a_n is the acceleration, which the accelerometer senses. That is, different angles correspond to different accelerations.

Figure 11.3-10 Linear acceleration simulator

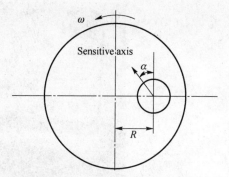

Figure 11.3-11 Rotation angle of the follow-up table

11.3.4 Simulation Computer

For simulation computer, the general computer or special computer may be used. In general, high configurations are required. CPU should possess high speed computing capability and enough interfaces such as A/D, D/A, SIO, DI, and DO. Real-time operation system is normally used in simulation computer.

11.3.5 Infrared Target Simulator

Here, only discuss dynamic IR scene projection based on the digital micro-mirror device (DMD). The dynamic IR scene projection system is composed of a graphics computer, a driver, an IR image converter, and an IR optically coupled system. See Figure 11.3-12.

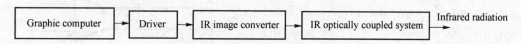

Figure 11.3-12 Dynamic IR scene projection

The graphics computer generates IR image of target and background. The driver receiving infrared image drives the infrared image converter (DMD) to work. The infrared image converter radiates infrared wave. The optically coupled system transmits infrared wave to an infrared seeker. Figure 11.3-13 indicates an infrared target simulator.

The technique specifications of the infrared target simulator include:

Figure 11.3 – 13   Infrared target simulator

- work wave band   3~5 μm,
- image resolution   >512 lines,
- image contrast   ≥0.22,
- no-uniformity   <5%,
- geometry aberrance   ≤5%,
- input signal   PAL.

The DMD is developed by Texas Instruments Company. It has a lot of metal micro reflection mirrors (13.68 μm×13.68 μm). Each micro-mirror may be controlled to deflect so as to switch on/off the reflection mirror. One of the micro-mirror plane may reflect light, but the other may not reflect light. When a micro-mirror is switched on, the reflection light can go through the projection lens. Contrarily, no reflection light can pass the projection lens. The imaging principle is shown in Figure 11.3 – 14.

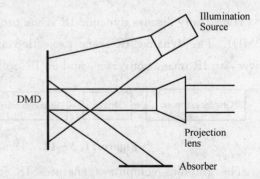

Figure 11.3 – 14   DMD image generation

In essence, dynamical images are obtained based on reflection modulation principle. As special wave band light source is used to illuminate DMD, each micro reflection mirror generates a pixel, and a lot of pixels generate an image. On-time length of a micro-mirror in each frame represents greyscale levels.